TURF MAINTENANCE FACILITY DESIGN AND MANAGEMENT

TURF MAINTENANCE FACILITY DESIGN AND MANAGEMENT

A Guide to Shop Organization, Equipment, and Preventive Maintenance for Golf and Sports Facilities

John R. Piersol
Harry V. Smith

WILEY

John Wiley & Sons, Inc.

Published by John Wiley & Sons, Inc., Hoboken, New Jersey
Published simultaneously in Canada

For general information about our other products and services, please contact our Customer Care Department within the United States at (800) 762-2974, outside the United States at (317) 572-3993 or fax (317) 572-4002.

Wiley also publishes its books in a variety of electronic formats. Some content that appears in print may not be available in electronic books. For more information about Wiley products, visit our web site at www.wiley.com.

Library of Congress Cataloging-in-Publication Data:

Piersol, John R., 1948-
    Turf maintenance facility design and management : a guide to shop organization, equipment and preventive maintenance for golf and sports facilities / John R. Piersol, Harry V. Smith.
      p. cm.
    Includes index.
    ISBN 978-0-470-08105-1 (cloth)
      1. Golf courses–Maintenance. 2. Turf management. I. Smith, Harry V., 1945-
II. Title.
  GV975.P54 2008
  796.35206'8—dc22

                                    2008011007

Printed in the United States of America

10 9 8 7 6 5 4 3 2 1

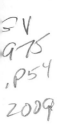

Contents

Introduction

It was common in past years to hear golf course maintenance facilities referred to as "the barn." There often was good reason for such terminology as golf courses were, and still are, often carved out of farmland sold for development. Farm buildings often remained on the property, and the barn was often converted into the golf course maintenance building used to house all the turf equipment. The barn kept the equipment dry, but it often had a dirt floor and inadequate space for organized equipment storage, for an area for repairs, and for an organized parts room.

Over the last 30 years, golf on television has drawn more attention to highly manicured turf, and many golfers began to demand high turf maintenance at their own clubs. This put pressure on the golf course superintendent to maintain the tees, fairways, and greens at higher standards, and the superintendents put pressure on the turf equipment manufacturers to produce more sophisticated mowing and cultivation equipment that would produce the type of turf golfers were demanding. More sophisticated equipment decked with hydraulics, electronics, and eventually some computerized components all translated into higher equipment costs.

As equipment budgets grew, golf course superintendents began to realize that jamming all this expensive turf equipment into a barn was not a good equipment management practice. Space was needed to store the equipment in an organized fashion so that the equipment technician and operators could easily access all units. Adequate space was necessary for preventive maintenance procedures and repairs. An organized parts room became more important. Cleanliness became important as it made no sense to store expensive equipment in a dirty environment.

The golf course mechanic began to evolve into the turf equipment technician as the technician had to constantly update his or her skills to stay current with new equipment technologies. Overall maintenance facility design and organization became more important in order to have a clean, efficient shop that allowed the technician to properly maintain the equipment to high standards. Good preventive maintenance (PM) translated into better equipment performance, longer equipment life, less unproductive time, better residual value when old equipment was traded for new units, and happier and more productive employees.

The golf course superintendent began to request a more highly trained technician rather than someone who was just good mechanically. The superintendent wanted someone who could help design a shop, organize a shop, develop and properly stock a parts room, establish PM programs for each piece of equipment, use a computer, help train equipment operators, and keep updated on quickly changing equipment technology. Such technicians were hard to find as very few schools offered specialized turf equipment programs because this specialty in mechanics was not widely known. The demand for the turf equipment specialist continued to grow.

Maintenance facility design and management for all sports turf operations is critical. Expensive turf care equipment needs to be washed after each use and stored in a dry, clean environment. Clean equipment allows for quick visual inspections to pinpoint any problem areas, and a clean floor becomes part of the PM program as a drop of oil under a mower can be easily seen and may be an indication of a hydraulic line that needs repair. A clean shop is an efficient and safe shop, which translates into cost savings. A clean, well-organized shop also builds pride in the employees, making them more productive.

The entire concept of the turf maintenance facility and the turf equipment technician is evolving. The maintenance facility is the epicenter of activity for everyone involved in turf maintenance at golf courses and sports facilities. Environmental and government safety regulations, equipment costs, increasing salaries, and the need to build a loyal, dependable team of employees have all had an impact on maintenance facility design.

The location on the property of the turf maintenance facility, the design and organization of the interior space of the facility, and the development and implementation of a PM program for the turf care equipment are all critical to the establishment of a cost-effective, efficient turf maintenance program. Many golf course superintendents and sports turf managers have developed ingenious designs for shop areas through years of experimentation and by working with golf course architects, turf equipment technicians, and other professional turf managers; however, there is no one source one can use to gain insight on efficient turf maintenance facility design and management.

The authors have compiled photographs, drawings, and concepts about turf maintenance facility design and management that will be a good source of information for the golf course superintendent or for the sports turf manager who is designing a new facility or who wants to better organize an existing one. Having all this information in one book will provide a convenient source of ideas and information that should stimulate the turf professional to think how to best design and implement programs that best meet the needs at his or her facility. No book has all the ideas, but the successful, proven concepts provided here should aid the turf professional in developing an efficient and safe turf maintenance complex.

The Role of the Turf Equipment Manager

History
The Management Team
The Turf Equipment Manager as the Team Psychologist
Future Roles
Sports Turf Facilities

HISTORY

Lake City Community College (FL) started its Turf Equipment Management Program in 1973. From the beginning, the turf equipment program was unique because it dealt with mechanics as it pertained to the specialized turf care equipment at golf courses. There were, and still are, very few such programs. Lake City Community College initiated its program as a result of feedback from golf course superintendents who indicated a desire to hire mechanics with specialized knowledge of turf care equipment repair, preventive maintenance (PM), and reel grinding skill.

In the mid-1970s, the turf equipment manager was still commonly referred to as the mechanic or perhaps the golf course equipment mechanic. The mechanic at that time was usually someone who had a good mechanical background, possibly from automotive mechanics or from the military, and who was very good at fixing equipment and doing basic welding. Reel mower maintenance was learned on the job, and the good mechanics learned quickly. Many of these mechanics were weak in implementing shop design and shop organization, reading equipment manuals, establishing PM procedures, setting up a parts room, and keeping a proper parts inventory. Mechanics from the military were desirable as they usually had strong organizational skills and a PM background.

Most golf courses and sports facilities got along satisfactorily with these early mechanics because they were very good at keeping the equipment running and at fixing anything that broke, and that is mainly what they were asked to do. After all, the equipment was stored in the barn where the mechanic worked, and there was not strong emphasis on shop organization, neatness, and PM.

From the mid- to late 1970s, golf on television grew more popular and as a result, millions of people saw highly manicured turf. This created a desire in many golfers to have their home courses more closely groomed. This desire translated into pressure on golf course superintendents to initiate higher levels of turf management. Even though new turf varieties, fertilizer products, pesticides, etc., were all involved in this movement toward highly refined turf management, it was the mowing and renovation equipment that had the most immediate impact, and the turf equipment manufacturers responded to the needs of the superintendent.

The turf equipment became more sophisticated; more hydraulics, electrical components, and eventually computerized parts became standard features, requiring a technician with more technical knowledge to be able to read and understand the maintenance manuals. Almost overnight, it seemed, golf course superintendents began requesting skilled turf equipment technicians to maintain the newer, more technical equipment.

This was the beginning of the turf equipment technician shortage. Demand for properly trained technicians was up, but there was no new supply coming into the market. The few turf equipment students that Lake City Community College graduated were readily hired by golf course superintendents. With direct input from turf equipment manufacturers and others in the golf industry, the Lake City curriculum was developed to include not only welding and mechanics, but also shop design, shop organization, hydraulics and electrical systems, parts inventory management, PM concepts, reel technology, and computers. These were the skills and training the golf course superintendents wanted in a technician, and a serious student could be satisfactorily trained in less than a year.

Through the 1980s and 1990s, the demand for turf equipment technicians continued to grow, and the supply of graduates from programs remained low with a resultant increase in salaries. Instructors and administrators at the few turf equipment programs that did exist were all trying to increase student numbers, which proved difficult. It would seem logical that with high industry demand and good, increasing salaries, it would be easy to recruit students for the programs. This was not the case, primarily because of total lack of career awareness. Nobody knew what a turf equipment technician did, and most people had no idea that a golf course even needed a mechanical person who could manage a shop with a million dollars or more in turf care equipment. How would people know? Who would tell them? Even golfers and green committee members had little knowledge of what went on in a turf maintenance facility.

The inability to attract students to turf equipment programs forced some programs to close. This was unfortunate as the faculty and staff at schools were trying to meet an industry demand and trying to interest students into a lucrative career. More programs were needed, so it was especially harmful when programs had to close because of low enrollment.

Lack of career awareness, low enrollment in turf equipment programs, and high industry demand for technicians still plague the golf and sports turf industries today. The equipment continues to get more sophisticated and expensive, and demand is growing for a turf equipment manager, not just an equipment technician. Salaries are very good, and jobs are plentiful. At Lake City Community College, it is common for the school to receive 30–50 job offers for 10–15 turf equipment graduates. The starting salaries offered range from $23,000 to $40,000, with career potential for $50,000 to more than $70,000.

THE MANAGEMENT TEAM

The golf course management team used to be the golf course superintendent and the assistant golf course superintendent, but now the turf equipment manager is included as an integral team member. In fact, most golf course superintendents will quickly declare that the turf equipment manager is their critical team member. It takes a huge load off the superintendent and the assistant to have a mechanical, management-oriented equipment manager who keeps the shop neat, clean, and organized, and who can implement a PM program for all the equipment so that everything runs properly. This allows the superintendent and the assistant to concentrate on agronomic practices, leaving the equipment and shop management to the turf equipment manager.

The team approach to management is important, as it takes various talents to run an effective turf management program, and everyone must be appreciated for his or her contribution. This is certainly true with the turf equipment manager.

The days of treating the equipment manager as "just the mechanic" are long gone. The equipment manager can keep the shop safe and efficient, save the maintenance budget thousands of dollars through proper inventory control and PM, and maintain high equipment trade-in value through proper maintenance practices.

THE TURF EQUIPMENT MANAGER AS THE TEAM PSYCHOLOGIST

A clean, organized maintenance facility has an impact on the psyche of the crew. It is common for people to act according to the condition of the space within which they live or work. If people work in a pigpen, they usually act like pigs, but if their work space is clean and neat, they will again act accordingly. So the way the equipment manager keeps the shop area affects the attitude and behavior of the crew; thus, the label of "team psychologist" for the equipment manager.

In a well-run facility, the entrance road is paved and leads to a neatly paved crew parking area complete with lined parking spaces. The outside of the building is kept painted and clean and is finished with appropriate landscaping. The interior spaces are all kept clean and organized from the reception area to the hallways, to the offices, to the crew lounge, to the bathrooms and locker areas, and to the shop. This neatness sends a definite signal of pride and respect for all who work there. For a prospective employee coming for an interview, a strong statement is made without a word being spoken.

Compare the previous situation to a maintenance facility where the entrance road is a dirt road riddled with potholes that leads to a dirt or dirt and gravel area where people can park as they see fit. The outside of the facility needs paint and is dirty, and the interior spaces are not well organized and are not kept very clean. The bathrooms and locker areas are not clean, and the shop area has a dirty floor, is poorly lit, and is unorganized. If you were coming for an interview, would you want to work here? What if you interview at the clean, organized facility and here, and the superintendent of the dirty facility offers more per hour; would it be worth it? Maybe some would opt for the higher starting wage, but one wonders how long the new hire would last.

Thus, the team psychologist, the turf equipment manager, can affect who wants to interview and crew turnover just by helping the superintendent keep the facility clean, neat, and organized. This is another way that a management-oriented equipment manager can save money. Most people want to work in an area where they feel good and where they sense that management really cares about their well-being. The turf equipment manager, working with the superintendent, can create this positive work environment.

FUTURE ROLES

It is difficult for an individual golf course superintendent to locate a skilled technician for an 18-hole course, but there is increasing demand for technicians to move into more management-only roles at multicourse facilities and with golf management companies that oversee many golf courses. This has occurred as a natural progression in the golf industry.

Multicourse facilities might have four or more technicians, creating an obvious need for a lead technician or a head turf equipment manager. There is a need for one person to be in charge of the shop area, to communicate with the golf course superintendent, and to train and supervise the other technicians. The titles *head equipment technician* and *turf equipment manager* are used interchangeably in the industry and both are respectable titles; however, *turf equipment manager* usually more clearly conveys a management-level position to those who do not really understand the job.

Job descriptions written by golf course superintendents state "head equipment technician" or "turf equipment manager" when they are looking for a lead person to take on the responsibility of overall equipment and shop management. This person needs to know how to design a shop, organize a shop, set up a parts room, order parts properly and keep a proper parts inventory, read all the equipment manuals to set up proper preventive maintenance schedules that follow the manufacturers' guidelines, use a computer and equipment management software, implement PM and repair procedures, and train new technicians. It is definitely a management role, and it is a difficult position to fill. A person with some formal mechanical education and training, years of experience with turf equipment, and some management experience is ideal for the turf equipment manager position. That is easy to say, but such a person is difficult to find.

With the growth of management companies overseeing the entire turf management at various golf courses under contract, demand is being created for a management-only turf equipment manager. If a company has ten golf courses under management, for example, a need will quickly develop for someone to supervise all the shops and coordinate turf equipment operations. Companies increasingly recognize that there would be cost savings and better management control if equipment, parts, and supplies purchasing were centralized; if the shops under contract were all organized in a similar layout; if PM procedures were standardized; and if common forms were used for all reporting. This type of coordinator developed on the agronomic side, but it took a while for the need for standard operating procedures for equipment and shop management to be recognized.

At first, some management companies let the golf course superintendent and the turf equipment manager operate the shop as a separate entity at each course. After all, that is what happens at individual golf courses. However, once the need for standard operating procedures was realized, this created a demand for a person who was an experienced turf equipment technician with a strong management background. This is a management position that requires the person to do hands-on mechanics only to help train someone at a site.

This multicourse turf equipment manager is another management step up from the head technician or turf equipment manager at an individual 18-hole or multicourse facility. This is an even harder person to find than the equipment manager at a golf course. The source for such managers is practicing equipment managers who have excellent mechanical and organizational skills, who are good communicators, and who deal well with a corporate structure. Because of the low supply of such specialists, salaries are very negotiable.

Future roles of the turf equipment manager will require stronger communication and management skills in combination with mechanics. Understanding budgets, handling personnel management, and developing standard operating procedures that make the shop a neat, clean, safe, and efficient operation will be increasingly important. Some golf course superintendents have had to use the titles *head equipment technician* and *second golf course superintendent* to justify higher pay for the equipment manager to upper management in order to attract qualified applicants. Through education on the management role that the turf equipment manager plays in the overall operation of a golf course, this dual title should no longer be necessary in the future. The title *turf equipment manager* should clearly indicate a management-level position requiring an appropriate pay scale.

SPORTS TURF FACILITIES

The previous examples in this chapter refer to golf courses, but sports turf facilities are in a similar situation. Most of the turf care equipment used in the sports turf industry and at golf courses is the same, and there is a similar need to have neat, organized, efficient shops.

Not all sports turf facilities have an equipment technician who works on equipment only. Many sports turf equipment technicians also do field operations or are irrigation technicians as well. Budgets commonly dictate the role of the equipment technician in sports turf, but increasing equipment costs and the need for cost-effective shop operations will continue to put more focus on the role of the equipment technician.

The information on golf course turf maintenance facilities and on the role of the golf course turf equipment manager adapts easily to the sports turf industry. The authors believe there is relevance in this chapter and throughout the book for the sports turf manager.

2

Maintenance Facility and Shop Design

Site Selection
Sample Layouts
Design, Construction, and Selling Your Design to Management
Remodeling Older Facilities
Meeting Regulations and Zoning
How to Allow for Expansion
Functional Interior Space
Zonal Concept
Equipment Storage Area
Office Space
Lounge and Break Rooms
Parking
Landscaping
How the Maintenance Facility Sets the Example

SITE SELECTION

Site selection is a critical decision for an efficient maintenance facility. There are usually several parties competing to locate the site—often the club pro, the director of golf, the greens committee, or, with a sports turf facility, the county manager, an architect, or someone allied with the turf being maintained. Each has a preference for the maintenance facility location. Sometimes, the desired site is not an effective location. The "we do not want to know it is there" or "keep it out of sight" mentality often prevails. Sometimes, these factions do not understand that careful siting, grading, and landscaping can produce an eye-pleasing facility in an efficient location. They want the facility out of sight and far away. Some facilities can be found completely off the main property. This can result in permanent, unnecessary expenses. The cost of long traverses is well known. There is increased fuel usage, as well as costly additional man- and machine hours. These extreme distances become tyrannical. Communication is difficult, and rapid response to emergency situations is impaired. If you have the opportunity to participate in the site selection for your facility, be sure to keep this in mind.

Other site selection criteria include access for semi-trailers (including sufficient space for turning and backing), access for employees and their vehicles, and the ability to adjust the site orientation for weather considerations. For example, a facility that has the shop area facing south in the southeastern United States will have to cope with extreme heat loads during the summer months. One facility in the southeastern United States with a southern exposure consistently had shop temperatures above 105°F by 1:00 P.M. each day through most of July and August. This necessitated changes in work schedules that were disruptive and expensive. It was simply too hot to work efficiently in the shop after noon. During the site selection and planning phase, this building could have been turned to orient the equipment storage area toward the south as a buffer and prevented this problem.

Selection Tip

When locating an 18-hole golf course maintenance facility, an accessible spot between the nines is ideal. With 36 holes, place the facility between the courses when possible.

Facilities with drive-through layouts have excellent accessibility for equipment and material deliveries but can pose a potential security problem because of the necessity for two gates or entranceways. These layouts also require clear signage to enforce a one-way traffic pattern to prevent collisions.

Dirt berms are an excellent way to reduce the visibility of the facility. The added benefit over fencing and plantings is reduction in noise transmission to the surrounding areas.

SAMPLE LAYOUTS

The layout depicted in Figure 2-1 represents a lot of careful thought. A standard pre-engineered metal building footprint was used for this design. Stephen Tucker, the equipment manager at the Ritz-Carlton® Members Club, created this design. It is the result of his many years of experience with shop layouts. He has designed or remodeled four facilities and consulted in the design or remodeling of several others. His use of a jib crane for heavy lifting, use of pegboard for the walls, and the location of the equipment lift in front of a roll-up door are efficient ideas, though certainly not unique to his designs.

The shop walls are 3/4" plywood furred with 5/8" strips, then covered with pegboard. The pegboard has rows of nominal 1/4" holes that offer great flexibility in hanging objects in a variety of configurations. Additionally, pegboard gives the walls a uniform appearance, protects the underlying insulation, and offers sound attenuation.

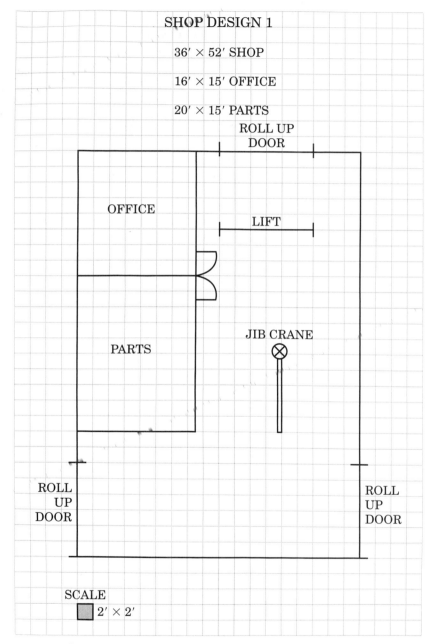

Figure 2-1   This layout was designed by equipment manager Stephen Tucker at the Ritz-Carlton® Members Club in Sarasota, Florida.

Figure 2-2　A chain link and slat wall hand tool storage system at the Ritz-Carlton® Members Club in Sarasota, Florida, uses space efficiently.

Another design element that is efficient for hand tool storage is a chain link enclosure, shown in Figure 2-2. Equipment can be hung on both sides if necessary, and secure storage can be created with a locking gate.

Other design considerations are painted and color-coded floors, light gray or white paint wall schemes to reflect light toward work areas, and employing skylights or solar tubes to improve light levels economically.

The shop design shown in Figure 2-3 is the very familiar big-box building. This features an open shop layout with the equipment storage and shop in a common area. This is an example of adapting to a preexisting design. A partial physical barrier is created with shelving and floor striping. This is not the ideal shop, but the design is well adapted to the space.

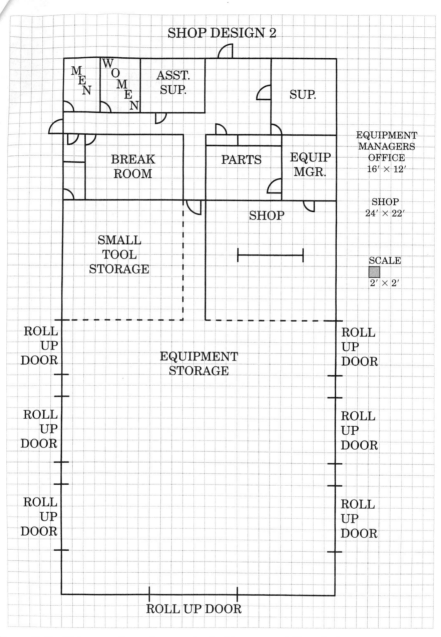

Figure 2-3   This open layout was adapted to a preexisting design.

The shop design shown in Figure 2-4 is another big-box pre-engineered metal building, but the shop is physically separated from the equipment storage by the administrative area. The equipment manager who devised this layout felt that moving his office and parts storage to the administrative section of the building would be advantageous. He feels he is only slightly cramped for space. This design seems to be working efficiently. He regrets that he cannot employ a jib crane with the present layout, but moving the office would allow a crane to be effective. Whenever you consider a jib crane, always be certain it can swing a full 360 degrees to make it useful.

The shop design shown in Figure 2-5 is a spacious layout for a 36-hole private golf course. Only walk mowers are stored in this shop overnight. All the other equipment resides in a multi-door storage area on the other end of the building. A jib crane is utilized efficiently. This shop was set up for a technician in a wheelchair and provides a number of innovative features.

DESIGN, CONSTRUCTION, AND SELLING YOUR DESIGN TO MANAGEMENT

There are a number of design considerations. The building construction budget is usually the limiting factor. The type of building is dictated by budget, local zoning, and visibility. If you have to match or blend with an existing building, then building costs are generally higher. The majority of maintenance facilities utilize pre-engineered and prefabricated metal building systems. These buildings offer low per square foot cost, rapid installation, and simple expansion capability. They are not suitable for every site but offer great flexibility for future expansion. American Steel, Butler, Gulf States, Star Building Systems, and Alliance Steel are just a few of the steel building manufacturers. A more exhaustive list can be found at the Metal Building Manufacturers Association website, http://www.mbma.com/display.cfm?p=41&pp=4

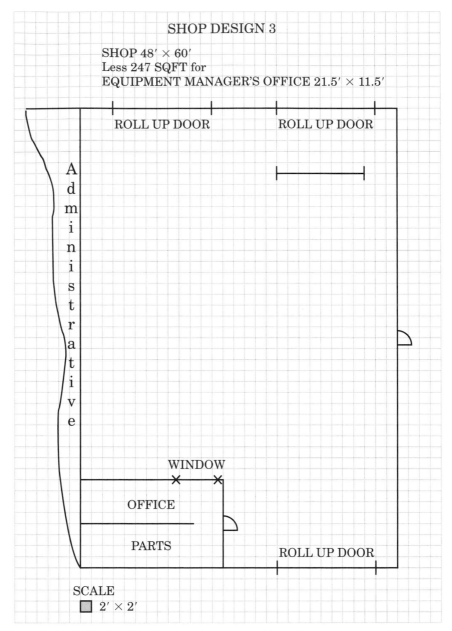

Figure 2-4   In this layout, the shop is separated from the equipment storage by the administrative office.

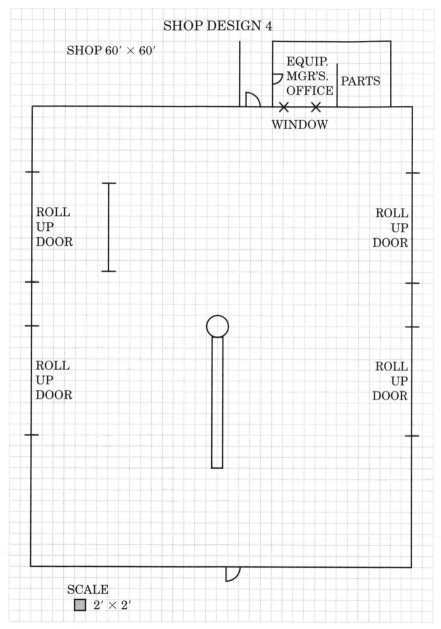

SHOP DESIGN 4

SHOP 60′ × 60′

EQUIP. MGR'S. OFFICE

PARTS

WINDOW

ROLL UP DOOR

ROLL UP DOOR

ROLL UP DOOR

ROLL UP DOOR

SCALE
2′ × 2′

Figure 2-5   This spacious layout was originally set up for a technician in a wheelchair.

There are many other materials available for turf maintenance facilities. Concrete block and bar joist, traditional wood frame, and concrete panel are just a few of the alternatives. Some building codes require special weather, wind, or earthquake resistance, which in turn dictates the material choices.

Selling your design considerations or getting management to accept your input on the facility can be difficult. Ultimately, the staff that works in a particular area needs to be involved in the design of that area. If the spray technician is not consulted about the chemical building or the equipment manager is not consulted about the shop, then valuable input is lost. The superintendent or facility manager needs to sell the expertise of his or her staff to management and to the architect. Good architects are highly skilled in many areas, but most have never had to work in a shop or chemical storage area they designed. It is necessary to emphasize this convincingly to management and the design team.

Building Design Tip

Insulated concrete form (ICF) technology is becoming a viable alternative building material for maintenance facilities. Costs are slightly higher than those of metal buildings, but there are several desirable paybacks. ICFs are very energy efficient, which is appealing in colder climates and also where designs include air conditioning for the shop space. They are very quiet, which can be a major concern when there is administrative space near running equipment. Another plus is the wind resistance of this design. The poured-in-place concrete walls are tied to the floor slab, and when the roof is properly designed, the entire structure can have as much as a 170 miles per hour wind rating.

The following are some design suggestions gleaned from interviews and observations at numerous existing facilities:

Air compressors can be a disturbing source of noise and heat. There are two ways to solve this problem: either mount them outside the building and away from the shop and administrative area, or be prepared to spend additional dollars to install a much quieter, rotary screw-type compressor. You will need to budget nearly twice as much for a rotary compressor as you would for a piston-type compressor if you want a quiet unit. Be certain to size your compressor correctly. Most of the industrial supply companies that sell compressors have charts and formulas for ascertaining the correct capacity for your application. An undersized compressor will run almost continuously, be short lived, and consume a lot of power.

Another area of potential aggravation is foot traffic patterns. Many designers fail to analyze traffic and work patterns in the shop. If the crew is routed through the shop to the break room, as shown in Figure 2-6, then shop efficiency is affected. This unnecessary traffic is both a safety hazard and a perpetual distraction to the technicians.

Another common practice is using the shop repair and maintenance area for unit storage. On the surface, this seems like a good idea because it provides a maximum utilization of space. In reality, the technicians are constantly shuffling equipment around to have work room. This is a huge waste of man-hours. If there is inclement weather, they either have to move the equipment out into the weather or try to cram it in a corner and work around it. Since the equipment is stored in the shop, the crew must come into the shop to take out or return their equipment. This is a distraction and a safety hazard. Unless there are horrific budget constraints, it is very desirable to provide separate storage for equipment. This will help maintain maximum efficiency and safety in the shop.

Figure 2-6   In this shop, all foot traffic to the break room goes through the shop. This door leads from the shop to the break room.

Some facilities have been designed with ten-foot eave heights. The logic behind this choice is usually visibility and/or cost. The actual cost differences between a 10- and a 12- or 14-foot eave height are minor. The cost difference may be as little as five percent. The advantages of the 12-foot eaves are more room for the equipment lift, more under-lift clearance with equipment on the lift, and better ventilation. With the 14-foot eave, you have the additional capability of adding a mezzanine for storage.

REMODELING OLDER FACILITIES

Remodeling an older facility will yield an immediate return on your investment with a few simple steps. It is often not necessary to tear out walls or add square footage. Many facilities can benefit from some basic reorganization and cleanup. One of the most dramatic transformations of a maintenance facility was accomplished with paint and elbow grease. An older shop in South Florida was a cluttered, oily mess. The employees could no longer park within the maintenance facility fenced parking area because of the accumulation of junk equipment in it. The interiors of the shop and equipment storage area were dark and filthy. Diesel soot and dirt had accumulated on the fluorescent light tubes, greatly reducing visibility. One storage room was so packed with blowers, hover mowers, and other junked handheld equipment that it could no longer be entered safely. The equipment manager filled a large open top trash container twice with junk, then found and resuscitated several pieces of equipment. Next, he reclaimed the parking lot for the employees. Once the employees were parked within the facility, the course owner noticed the absence of vehicles outside the fence and surmised no one had showed up for work. He got a nice surprise when he came to investigate. The superintendent was ambivalent about the equipment manager's cleanup activities until he received several compliments from other superintendents who came to visit. Once the superintendent saw the advantages of a clean, neat shop, he became an energetic supporter of the equipment manager's efforts. Eventually, the floors and walls were painted, an equipment lift installed, and the light fixtures cleaned. The difference was astounding. Even the crew adopted an improved attitude about their facility and their equipment.

Other remodeling efforts may not be as inexpensive but can certainly be rewarding. If an older facility has insufficient lighting, ventilation, or shop space, then a major remodeling is in order. Frequently, lighting levels were designed for storage space, not workshop space. Often, a higher lighting level can be obtained from modern bulbs and fixtures and still yield a reduction in overall electricity consumption.

Figure 2-7   The Country Club at Boca Raton in Boca Raton, Florida, an older shop, was dramatically improved by cleaning and painting.

Insufficient lighting levels waste man-hours and electricity, as the technicians always have to supplement the work area lighting with portable lamps. Lower light levels can mean eyestrain and overlooked faults when servicing equipment.

Equipment storage woes can be addressed in a number of ways. A mezzanine, as previously mentioned, can be installed to increase storage space for lightweight items, but be careful about floor loading and ease of access. Another simple trick is to measure the available equipment storage space, draw the layout on graph paper, measure each piece of equipment's footprint, transfer the measurements to cardboard cutouts, and start experimenting with parking patterns. No equipment has to be moved until you have a workable pattern. If the new pattern is workable, then line off and number each parking space to correspond to the machine identification. A bonus to this method is that because every machine is parked in the same place every night, any fluid leaks can be traced to a specific machine.

Shop space can also be addressed by additions or extensions to the existing space. Carefully assess the accessibility of any additions. A shop space that is difficult to drive into or exit will go unused. Sometimes, a fresh start is in order. One facility had several small buildings scattered throughout the maintenance compound. Some had crude breezeways between them. After careful examination of the existing facility, the decision was made to locate a new shop and storage area across the street. Rehabilitating the old structure was not as cost-effective as replacement.

MEETING REGULATIONS AND ZONING

Facility planning can be a real headache if you are not well versed in the zoning laws and regulations applicable to your location. This tends to be a little less intimidating when the facility is part of a larger project, such as a sports complex or golf course, but no one escapes the pressure of these regulations. You will find that hiring a local contractor who has a successful track record with the local planning and zoning boards can save you time and money. There are numerous incidences of zoning and permitting taking an extended period to complete after the rest of the development is finished. One golf course operated its maintenance facility out of two tents, a trailer, and several cargo containers for more than 18 months after the course was grown in and operating. In this case, there were zoning problems plus water use and environmental permitting delays centered around the wash rack area. Budget freezes are also common, and often a cost overrun somewhere else in the project siphons funds from the maintenance facility. It is wise to stay in constant communication with the planning and zoning boards governing your project. Always be alert to the impact of change orders on the permitting process. The placement of a roll-off waste container at a new maintenance facility resulted in the zoning board requiring a concrete slab and a concrete block wall to screen the receptacle from view. Simple changes or additions can dictate complex and expensive solutions.

HOW TO ALLOW FOR EXPANSION

Most equipment managers, superintendents, and turf equipment facility managers agree that they never have enough space in their facilities. Changes in agronomic practices, added turf maintenance responsibilities, and a myriad of other demands frequently leave the facility crammed to the rafters with equipment and short of shop space, equipment storage room, and administrative space. It is rare to be given the budget to build a new turf maintenance facility at an existing course or complex. It is much more likely that you will acquire a budget for remodeling or additions. Sometimes, the driving force is a state or a U.S. Environmental Protection Agency (EPA) regulation that cannot be met. There have been numerous older facilities that got much-needed refurbishing because of new regulations. An example of this was a 25-year-old facility that was under threat of hefty fines if the owners did not move their chemical storage and fertilizer storage from their main building. They built a freestanding, code-compliant chemical and fertilizer building with a chemical mixing room some 75 feet from the old building. This allowed the expansion and much-needed remodeling of the shop and storage areas into the old chemical and fertilizer storage area.

The lesson here is that the owners had never given any thought to future expansion or consideration of the radical changes brought about by changing needs and regulations. If you are constructing a new building, rest assured that in two or three years you will find need for additional space. With this in mind, plan for future expansion while the new facility is being designed. If at all possible, do not allow yourself to be literally painted into a corner on future expansion. There are a number of high-end courses that have nowhere to expand to meet their continuously increasing demand for more space. One complex had to locate an additional facility several blocks from the course at the edge of a residential neighborhood. The complex bought an existing piece of property and tore it down. The cost of the property and removal of the existing structure was more than the cost of the new building, and once again the complex gained no room for expansion.

Plan for expansion by building easily expanded buildings. Most pre-engineered metal buildings have wing and end panel expansion capability. Buying a building that is expandable and leaving room for it to be expanded is a low-cost option during the original construction.

FUNCTIONAL INTERIOR SPACE

Many new facilities are turned over to the superintendent or manager as large boxes with a few offices framed in for administrative purposes. How the interior shop space is to be utilized is left up to the occupants. There are many ways to divide the shop space, with the most basic method being tape or lines on the floor. In Figure 2-8, the equipment manager utilized a corner adjacent to the administrative area. He used floor tape, lubricant barrels, a shop-fabricated tubing "corner", and shelving to create and define his work space.

Another approach to separating the shop from the storage area is to line off the floor with tape and confine the equipment to one end of the building. If you have the luxury of parking your equipment two deep in front of each roll-up door and then creating a center aisle, you will greatly reduce time-wasting equipment shuffling.

Figure 2-8   This shop at Raptor Bay Golf Club in Bonito Springs, Florida, is built in the corner of a big-box building. The shop is conveniently located and out of the foot and equipment traffic patterns.

The layout in Figure 2-9 utilizes an aisle on one side of the storage building in an asymmetrical pattern. This was necessitated by the location of the administrative wing teed into the storage building where there would have normally been two roll-up doors in a center aisle configuration. The doors leading to the administrative wing are visible on the left.

As you separate and define your shop area, you will need parts storage space, a technician or equipment manager's office space, and the zones you wish to establish. Focus on safely separating incompatible zones. The battery zone needs to be well ventilated and away from sparks and open flames. This means keeping the welding zone and grinding zone at a safe distance. A good line of sight from the technician's office into the shop is desirable for safety and security.

Figure 2-9   This is the view from the equipment manager's office at the Old Collier Golf Club in Naples, Florida. Note that the shop, the fuel island, and part of the yard are visible through the window.

THE ZONAL CONCEPT

The zonal concept is not a new or unique organizational method. What is advantageous about it is the capability of placing each tool or operation in a designated space. The zonal concept is like coleslaw: categories can be chopped as fine or as coarse as you like. Use the fine-or-coarse approach as you see fit. Some facilities have zones that are divided into a few wide areas. For example, there would be an administrative zone, an equipment storage zone, a cleaning zone, a repair and preventive maintenance zone, a chemical storage zone, a fueling zone, and a bulk storage zone. Another facility might slice the zones much finer or even create a series of subzones. The administrative zone might be further divided into a zone for offices; a zone for the break room; a zone containing showers, restrooms, and lockers; and even a zone for the intern's living quarters. The shop area can be subdivided the same way, and in the Chapter 5 description of shop organization and suggested shop tools, you will see an extensive list of zones. This is an inexpensive method requiring only that thought be given to which zones are appropriate and what will be located within these zones. The zonal concept is not magic, just a handy way to organize.

The following list of zones was originally compiled nearly 20 years ago by the Georgia Golf Course Superintendents Association. Many zones have been consolidated at modern turf maintenance facilities, such as the fertilizer and seed storage areas, and some zones have almost disappeared; the paint zone and greenhouse are two examples. Many of these zones can be subdivided or combined to suit your purposes. This is a basic list to kick-start your thought processes about zones. Use this list to build your own custom zone list.

Turf Maintenance Facility Zones

Administrative Zone
Chemical Storage Zone
Equipment Maintenance and Repair Zone
Equipment Storage Zone
Fertilizer Zone
Fuel Storage
Golf Course Supplies
Greenhouse
Hand Tool Storage Zone
Irrigation Zone
Material Storage Zone
Paint Room
Parking
Seed Storage Zone

Soil Mix Storage Zone
Trash Collection Zone
Washing and Degreasing Zone

EQUIPMENT STORAGE AREA

The equipment storage area presents a number of challenges. You must decide which equipment must have direct access to the exits or create a design that allows access to all the equipment at any time. Full-time access to all equipment usually requires a center aisle and rows of roll-up doors on both sides of the building. A building or wing off the main building that is wide enough for two double rows of parking spaces on either side of a center aisle works well if the equipment on the outside row can exit through a roll-up door. Any equipment parked on the aisle can be pulled into the aisle, then out an exit. The efficiency rule for this layout is that no piece of equipment has to be moved to access another piece. If equipment is parked three, four, or five deep in a blind column, then accessing the equipment against the wall (first in) requires six, eight, or ten equipment moves to extract it. This is a waste of time and fuel and adds unproductive wear to the equipment. Cold starts and shutting off equipment before it warms up accelerate engine and hydraulic system wear.

As previously mentioned, locating the equipment in a separate area or building from the shop is an optimal solution. There are many issues that arise when equipment storage and shop space are shared.

OFFICE SPACE

As obvious as it may seem, office space should be predicated on staff size. If you have only a turf manager and an administrator, your office space is minimal. A 36-hole golf

Figure 2-10   The Cherokee Country Club in Atlanta, Georgia, is a good example of a well-lighted and neatly organized equipment storage area. Note the extensive use of floor striping as an organizational tool. Note also that no repair work is done in this storage area.

course will require a much larger staff and consequently more offices. Do not overlook security issues. A reception area that can be secured when no staff is in the building is imperative. There have been office equipment thefts when everyone is in the break room or away at lunch. A theft that is the result of an open door is hard to explain to the insurance carrier. Some facilities have chosen to secure their entrance with a magnetically locking door. If no one is in the reception area to buzz someone in, the door is locked. Be aware that there are back door problems, too. If someone can enter the office area through the shop or equipment storage area, then a locked front door does not help; there is also the danger of losing unsecured turf equipment. Because fairway mowers can cost more than $50,000 each, this can be a devastating loss.

Most equipment managers prefer to have their office in the shop area. A large window that overlooks the shop is desirable. This keeps the equipment manager apprised of activities in the shop while administrative duties are being performed.

The office size for the equipment manager should be adequate to comfortably arrange a desk, bookshelf, computer, and guest chair. Some equipment managers prefer to limit access and increase security by locating the entrance to the parts room within their office.

LOUNGE AND BREAK ROOMS

Break rooms have a huge influence on the morale of your employees. A dirty, cluttered break room sends a message to the employees that they are not valued. Beat-up coffeepots and dirty sinks and floors further reinforce this assumption. Remember that a clean, neat break room is not a costly budget item. A clean break room will tend to stay clean with very little additional supervision. Once the expected level of cleanliness is established, only an occasional reminder should be necessary.

Break rooms must be in an accessible spot in the complex. A break room that requires the crew to walk through the shop or administrative area will be a constant headache. You will have to deal with dirt, noise, and distractions to the employees along the path. A clever solution to the break room problems is a two-chambered design. This design was created and implemented by Tim Hiers at the Old Collier Golf Club in Naples, Florida. The entrance to the break room is adjacent to the equipment storage area. The crew comes out of the storage area and into the first chamber or lock.

This chamber contains soda and snack machines, a sink with prep counter, a refrigerator, and several microwave ovens. The floor is ceramic tile and covered with open grate-type matting to catch dirt and grass. If an employee is on a quick break, he or she does not have to enter the main break room. At lunch, employees get their lunches ready in the first chamber and then go into the break room to eat. The outer chamber acts as a thermal barrier, sound barrier, dirt barrier, and prep area. This reduces heat and air conditioning loads, keeps the main break room cleaner, and reduces food prep clutter in the break room. The break room also serves as a training room and is equipped with video players, monitors, a whiteboard, a clean uniform storage closet, and a media storage area.

Figure 2-11 In an innovative solution to a universal problem, each crew member at this facility has an assigned bag for his or her dirty uniforms.

Figure 2-12   Eating lunch in this neat, clean break room at the Ritz-Carlton®
Members Club in Sarasota, Florida, would be a pleasure.

PARKING

Parking areas can be either nearly invisible and trouble
free or one of those endless headaches. There are hundreds
of maintenance facilities that have parking problems. The
number one problem is insufficient parking spaces. Cars can
be found double parked, parked on the grass, parked in front
of the fuel pumps, and parked blocking the bulk storage bins.
All of this is unsafe, disruptive, and expensive.

Parking lots should be paved with clearly lined spaces.
Special parking spaces can be designated for handicapped
parking, guest, and other designations as required. If you
count the number of employees you have in your initial budget, add several spaces for your guests, designate spaces near
the office entrance for handicapped parking, as shown in
Figure 2-13, and add a few spaces for vendors and a few more

for good measure, you may stave off parking headaches for a few years. Parking is like building space: you will likely never have enough. Be sure to include an expansion plan for your parking area in your original design.

Please realize that the emphasis placed on parking is necessary because of the impact that is created by this "first impression" space. There is a message here for the employees as well as any visitor to the facility.

When planning a new facility, physically separating the parking area from the equipment storage area solves a number of potential problems. If there is a barrier of some kind between the two areas, then there are fewer problems with cars blocking bulk bins, equipment storage areas, or wash racks. If you can install a sign marked "No Personal Vehicles Beyond This Point" that incorporates a physical barrier, such as a gate,

Figure 2-13   This is a well-planned parking area with an inviting entrance. Note the building overhang that provides late-day shade for this facility at the Old Collier Golf Club in Naples, Florida.

then the troublesome mix of personal vehicles and turf equipment is avoided. It should also be obvious that providing enough parking spaces is mandatory for this to work. You will make your insurance carrier very happy if you can separate these two elements in your maintenance facility.

LANDSCAPING

Landscaping around the maintenance facility may seem like a waste of time and money. The actuality is that landscaping not only provides aesthetic qualities, contributes to that first impression, and can be a powerful tool for temperature moderation around the facility, but can also provide screening and noise abatement. Carefully chosen and strategically placed trees provide shade that is physiologically and psychologically beneficial. Trees provide summer shade that can reduce air-conditioning costs and reduce interior car temperatures. Observe any parking lot on a hot summer day. If there are a few spaces shaded by one tree in a vast parking lot, they will be occupied first. Providing shade for your parking spaces sends a message to your employees: "We thought of you when we designed this facility." Some facilities designate the shadiest parking spot in the lot for the employee of the month. The employee of the month concept is an inexpensive incentive that is well appreciated.

The placement of ornamental shrubs around the maintenance facility provides favorable aesthetics, as shown in Figures 2-14 and 2-15 and shrubs do add a slight insulating factor to building walls by slowing winter wind and by keeping the summer sun off part of the wall. The aesthetic qualities of landscaping around the maintenance facility should not be discounted. An attractive entrance road, a paved parking area, and a well-constructed and maintained maintenance building with landscaping all establish a positive

Figure 2-14   At the Golf Club of Georgia in Alpharetta, Georgia, the maintenance facility is artfully hidden by a berm and carefully selected greenery.

psyche for all who drive in. Attractive landscaping creates eye appeal from the road for a residence; it does the same at a workplace. Landscaping is not a necessity, but it supports a positive working environment.

HOW THE MAINTENANCE FACILITY SETS THE EXAMPLE

The appearance of the maintenance building speaks volumes about the overall facilities. A dirty, disorderly maintenance area says, "No one here cares how things look. They will not even keep their own backyard clean." This message is conveyed to the employees; an apathetic attitude like this is contagious and can infect the entire operation.

Most turf maintenance tasks require careful attention to detail on the part of the employee. It is important to observe how the mowers are cutting. Are there obstacles in the path

Figure 2-15   This is the back side of the berm shown in the preceding photo.

of the mower that may damage a reel? Is paper or other debris in the path of the mower? Will that debris be mowed into hundreds of pieces, making an unsightly mess? Is there oil on the turf, indicating a leak? It takes a lot of discipline on the part of the operator to make these observations when it is hot, when he or she has been mowing for hours and is mesmerized by the routine. Picking up trash requires stopping the mower, disengaging the reels, locking the brakes, getting off the mower, picking up the debris, and starting up again. So what does this have to do with a clean maintenance facility? Again, the atmosphere is set from the start. If the superintendent or sports turf manager is talking to the crew about being observant and picking up rather than mowing paper, for instance, he or she has more credibility when discussing this from a clean shop. It indicates to the employees that the supervisor really believes in what he or she is saying about the importance of neatness and provides an example by having a maintenance facility that makes that statement. It is easy to comprehend the dichotomy of a supervisor telling the crew to be neat and respectful out on the course or sports fields when the crew works out of a dirty, disorganized shop. The crew quickly notices such consistencies or inconsistencies and may react accordingly.

The maintenance facility does not have to be new or high budget to set the proper example. For instance, the turf equipment program at Lake City (Florida) Community College began in a weathered, unattractive wood frame structure with an old, pitted cement floor. The instructor taught the students to keep the facility neat, organized, and very clean. The program did not move into a modern building for almost 13 years, and the old facility continually showed the students and visitors that it was not necessary to have a new building to be neat and organized. This was an important message that was not overlooked!

The authors have given many talks to technician groups emphasizing the importance of shop management, and we frequently show a video of a new, upscale shop to demonstrate the neat and clean principle. Occasionally, someone in the audience chimes in that the facility obviously had a big budget, so it could afford such a sparkling shop. This always provides a perfect opportunity to point out that money is not necessary to be neat and organized, but usually the audience responds first, answering that the point was just to show an example of how important it is to have a clean shop. The rest of the audience gets the message!

Throughout the book, there are numerous examples of the neat and clean philosophy. The psychological impact on the crew and everyone involved with the operation should not be overlooked.

chapter 3

Preventive Maintenance

WHAT IS PM?

There is no universally accepted definition of *preventive maintenance* (PM), but the following definition will provide a good starting point: a planned approach to maintaining turf equipment designed to extend the life of the equipment, prevent catastrophic breakdowns, and ensure safe operation.

Note that there are three distinct elements: to extend the life of the equipment, to prevent breakdowns, and to make sure the equipment is safe to operate. Good PM will result in cost savings, time savings, and accident prevention. Any well-designed and properly managed PM program can be sold by emphasizing these three points.

PM has its modern roots in the Industrial Revolution, but its history goes back as far as the ancient Egyptians. Joel Levitt, in his book *Complete Guide to Predictive and*

Preventive Maintenance, says, "They used to do PM inspections on the great pyramids." PM is obviously not a new concept and yields tremendous cost savings when applied effectively. In addition to cost savings, PM balances workloads and yields predictable staffing requirements. Emergency repairs (unscheduled repairs) are minimized and overtime reduced.

SETTING UP A PM SYSTEM

PM systems do not have to be computerized, complicated, or expensive. There are many good PM systems that are inexpensive and paper-based. Several companies sell wall chart–based systems that do an excellent job of tracking and scheduling PM. Magnatag® has a paper-based system that tracks up to two years of PM on a single magnetic, wall-mounted board. There are PM systems that utilize nothing more than a large (4' × 8') enameled whiteboard. This is a very visual way to schedule PM, but it lacks a way to create permanent records unless someone is willing to record the board on a daily basis with a digital camera and archive the photos. Another problem inherent with any erasable board method is the onerous task of creating work orders by hand.

PM Plan Design: The Quick Start Method

Follow the instructions below to create a quickly implemented PM system on the fly. You will find that once a rudimentary system is in place, expanding and fine-tuning the system is much easier. Follow these first steps if you have no viable system. If you have a PM system, check it against these instructions to ensure you have the basic elements covered.

a. Create a Complete Inventory

Inventory all your equipment if you have not done so previously. If you have equipment without serial numbers, model plates, or other unique identification tags, create identity plates for these units. You need a complete inventory list not only for a comprehensive PM program, but also for insurance purposes, theft recovery, and depreciation analysis.

b. Assemble All Operator's Manuals

Gather all the operator's manuals for the equipment on your list. You may find that some manufacturers do not list all the PM tasks in the operator's manual. Consult the repair manual for these units. Most operator's manuals can be found on the manufacturers' websites. They are usually downloadable with few restrictions.

c. Extract Recommended Services

Read through the manuals and extract a list of all the recommended services and service intervals. Also check the service manuals to verify you have recorded all the services. Be aware that some services do not occur for the first time until the fourth year.

d. Create a Two-Year Service Chart

From your list of services, create a two-year service chart for each piece of equipment. This is not a comprehensive service chart. As previously mentioned, there may be services appearing after four years. You are interested in the immediate creation of a PM system, but remember that these later services must be addressed. If you are taking over an existing fleet and there are no PM records, you must begin service immediately using the triage method. Pick the most critical pieces of equipment, and immediately begin service on them. An easy way to assess the most critical pieces of equipment is to think of those units that could cost you your job if they fail.

e. Build a Log

Your list of PM services can now be created on graph paper, an Excel® spreadsheet, a legal tablet, a whiteboard (a high-quality whiteboard is usually enamel on steel and therefore compatible with magnets), a commercially available magnetic board, or a software-based inventory tracking program. If you do only three or four critical pieces of equipment initially, you are still headed in the right direction.

f. Start Now

The critical thing is to start now. If you began your PM system with the days method and you want to convert to hours (a more accurate and more frequently employed method), begin converting your units as they are serviced. Inexpensive, lithium battery–powered hour meters are available for any spark ignition engine. These units will run up to five years before they must be replaced. This means that something as small as a 22 cc weed trimmer engine can be equipped with an hour meter. A commonly accepted practice is to temporarily equip a backpack blower, weed trimmer, or other small engine–powered unit with an hour meter to establish a baseline for that class of equipment. If you observe that your hour meter–equipped weed trimmer is running 25 hours a week and there are several more units with the same duty cycle, it is a safe assumption that all units in this category should receive a 25-hour service every week. One or two inexpensive hour meters used judiciously can establish services for all your small handheld equipment.

Tips on Beginning a Preventive Maintenance Program

Many times, the creation of a Preventive Maintenance (PM) plan gets bogged down in the details. Typical hang-ups include the following: Should I build my schedule by days, weeks, and months? Should I build my schedule by hour meter readings? What if some of my units do not have hour meters? The fastest way to start a PM program is to use the calendar or day, week, and month method. Some of the advantages of this method are as follows: you can start immediately and not wait for a certain hour number to come up on the meter; no meters are necessary, so triage is immediate; and you can convert to the more accurate hour-based system at your leisure and when your budget allows.

Work Orders

Work orders are created as companions to the PM log. A work order is your paper (or computer) trail that verifies that your scheduled maintenance tasks have been performed. A record of all services performed on a specific machine has a multitude of uses. The work orders from two similar machines are valuable for comparing repair costs. The machine that has been costing more time and materials to maintain may be the one traded or sold first. Another subtle advantage is legal. If a machine is involved in an accident, the repair and PM records take on great significance. In one instance, an employee drove a utility car into a lake. The employee claimed the brakes failed. An examination of the braking system and the PM records indicated otherwise. The brakes had been checked and adjusted and the service logged on a work order the week before. The brakes were examined by an independent expert who confirmed they were still functioning correctly. A lengthy wrongful injury lawsuit was avoided.

In order to obtain and utilize this valuable information, a computerized system is your best alternative. With a computerized system you can not only track your PM, create work orders, and control inventory, but you can also do budgeting, plan effective replacement strategies, predict failures, and build amortization schedules.

Caution: The first step in creating a successful PM program is to get all the stakeholders committed to your effort. If upper management is not committed, you will fail because of funding and the insertion of other priorities ahead of your PM effort. If your technician(s) and equipment manager are not supportive of your efforts, your program will fail as a result of lack of effort and attention. Sell the advantages of this program up and down the organizational structure. Once you have convinced the appropriate people of the worth of your program, begin the design phase. Upper management, owners, and technicians must be convinced that PM will reduce their costs and make their jobs easier.

THE EFFECTS OF NO PM

The effects of no PM are many and varied. When you have no PM, chaos reigns. There is no way to accurately predict repair demands or repair costs. Staffing is impossible, and overtime is inevitable.

In spite of these facts, there are many turf maintenance facilities operating with no real PM program. In fact, it is likely that the majority of turf maintenance facilities have no true, comprehensive PM program. What are the results of this "run to failure" approach?

The biggest drawback to no PM is the chaos of unpredictable repair demands. There are no opportune times for something to break, but there are many very inopportune times. The day before a major tournament is one of the inopportune times when a critical piece of equipment might fail. It seems that most of the memorable failures occur at these disastrous times. This type of failure results in stress for everyone, from the technician to the greens committee. The superintendent, the technician, the director of golf, or the general manager experiences the most stress. Equipment failures have been known to cost jobs.

Another dilemma associated with unpredictable repair demands is staffing and overtime. How do you staff for the unknown? Typically, the reaction is to add staff to cover emergency repair operations. This leaves you overstaffed most of the time and perhaps still understaffed when disaster strikes. The next headache is overtime. If the unit breaks the day before it is mission critical to have it online, the technician(s) must spend the night in the shop. This is neither cost-effective nor morale-enhancing. No one wants to pull an all-nighter, especially on a regular basis.

The cost of overnight delivery of parts on a continual basis can become a cost factor. Parts purchased in conjunction with regular stock items and shipped the cheapest way are much more cost-efficient than one or two items ordered overnight.

PM does not promise less work. There will still be plenty of work to do. The promise of PM is a smooth, predictable workflow. Staffing becomes easier. Work orders are produced that are easy to stock for. You know what you will be doing tomorrow and the next day. Emergency or unscheduled repairs are reduced to a minimum. Catastrophic repair scenarios are minimized. Smooth, predictable workflow is a stress reducer and morale booster.

WELL-TRAINED OPERATORS AS A RESOURCE

A well-trained operator is one of the technician's greatest resources.

An alert operator who is not afraid to approach the technician with a potential equipment problem is an asset with a double value to the organization. He or she is valuable not only as an operator, but also as an experienced, additional set of eyes and ears for the technician. No one knows the equipment better than the operators. They live with the equipment. A technician may see only part of the equipment in use each day and then sees the equipment only briefly. A chafed hose or a bearing starting to chirp from lack of lubrication can be easily overlooked. A well-trained operator will notice a problem like this immediately. A corollary here is that the operator must be assigned the same piece of equipment every day. Operators who are continually rotated from fairway unit to fairway unit or triplex to triplex will not be as effective at spotting potential trouble. Familiarity with the sights and sounds of a specific piece of equipment yields early warnings of failure.

The only unsolicited testimonial these authors ever collected was in shameless praise of a PM-savvy technician at an older, private golf course in southern Florida. The operator had worked at a number of area courses over his 20-year career. He related that his habit for most of his career was to arrive 15 or 20 minutes ahead of the other crew members. This allowed him to select the "best of the worst" piece of equipment for the day's mowing. He was a conscientious operator and hated to operate a dirty, beat-up unit. He bragged that he no longer had to arrive early and pick through the equipment. He had his own piece of equipment, permanently assigned. The best part, he related, was that it worked beautifully and looked good, and he was proud to use it. What insight can be found in this comment!

The other obvious benefit of permanently assigning equipment is the technician's ability to spot a potential equipment abuser. One machine with an inordinate number of service issues means either an abuser or a lemon. No matter which condition is at fault, the information is valuable. The not-so-obvious result is that a competent, conscientious operator feels privileged and rewarded by clean, well-maintained equipment. The word goes out to the operator community when a course has superior equipment. You might dare to say that better operators are attracted to better equipment.

THE TECHNICIAN'S ROLE IN PM

The technician's role in PM may appear obvious. The technician executes or assigns the PM tasks, tracks the work orders, provides the quality control function for PM activities (and may execute all the tasks if budget and staffing so require), and updates or adds tasks as manufacturers change recommendations and as new equipment comes into the turf equipment inventory. Are these the only tasks the technician accomplishes? No. The technician, especially if positioned as an equipment manager or head technician, must continually "sell" the PM philosophy. Who must be sold? The management up the line and the employees down the line are the customers. The superintendent, assistant superintendent, general manager, fleet director, director of golf, or other managers must be reminded of the advantages of PM. PM programs often disappear with changes in management. For new general managers doing a cost analysis or inspecting the maintenance budget, the PM costs will jump off the page at them. The first thought when most managers encounter a large line item such as PM costs, is how to reduce or eliminate it. The equipment manager must be ever vigilant of potential budget cuts and prepared with a convincing presentation to support the PM costs. When new employees, especially operators, come on board, it is the responsibility of the equipment manager to train, educate, and convince them about the value of the PM concept. As mentioned previously, operators are the technician's early warning system for potential equipment problems. Operators must be convinced that PM is to their advantage and the equipment technician is their ally. If they fear retribution for reporting potential problems, the problems go unreported.

SOFTWARE

This is a very brief list of PM software. It is not intended to be comprehensive because the information management industry is changing so rapidly. You will find PM software if you do a web search under such topics as "fleet management software," "turf equipment management software," "preventive maintenance," and numerous other categories.

TRIMS

www.trims.com

TRIMS has been around since 1986. It is a wide and deep approach to managing turf equipment, parts, personnel, chemicals, and even tree inventories. TRIMS is bar code capable, which simplifies tracking equipment maintenance and management. TRIMS can generate any number of reports, work orders, or lists.

Bigfoot (formerly Smartware)

www.bigfootcmms.com

Bigfoot is a computerized maintenance management software (CMMS) package used for PM, work order management, inventory, predictive maintenance, asset tracking, labor, maintenance requests, tool crib, and more.

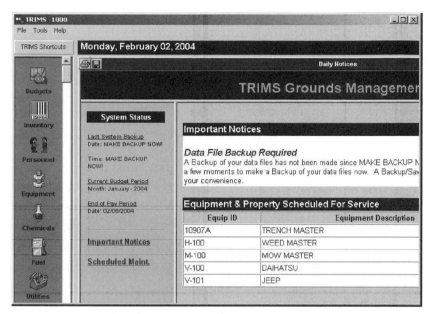

Figure 3-1   A screenshot from TRIMS software.

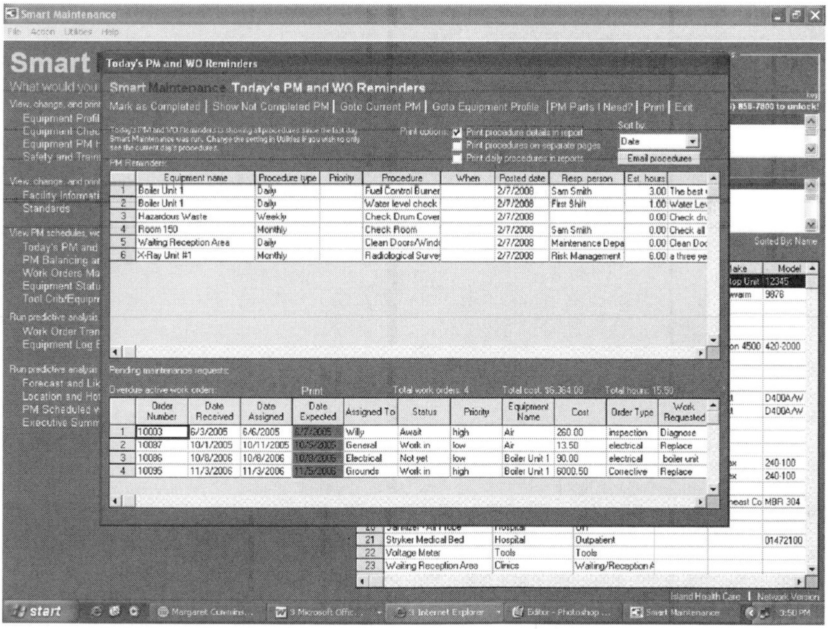

Figure 3-2   A screenshot from Bigfoot software.

Toro® myTurf™ Fleet Management

www.toro.com/golf/myturf/description.html

This program can track a fleet online, track and schedule PM, and track ownership costs.

TurfCentric® Software, SmartShop®, and GCS

http://www.turfcentric.com/gcs/products/smartshop.shtml

SmartShop® has a system similar to radio-frequency identification (RFID) technology that allows a module on each piece of equipment to transmit its identification number and hours of run time to a personal digital assistant (PDA) or desktop computer. This system can track maintenance intervals by miles, hours, or days. A bar code–capable system can also be employed. The GCS system includes equipment maintenance and a course management package for the golf course.

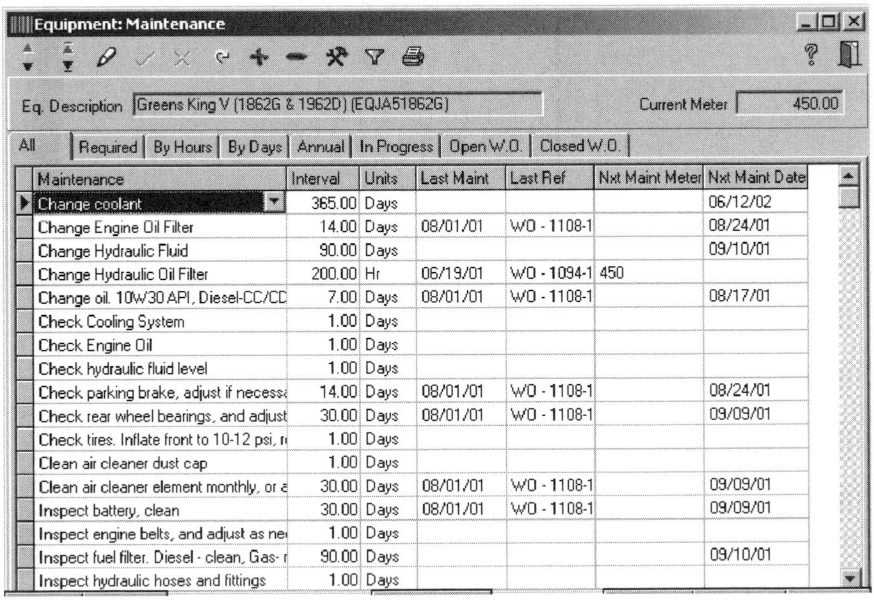

Maintenance	Interval	Units	Last Maint	Last Ref	Nxt Maint Meter	Nxt Maint Date
Change coolant	365.00	Days				06/12/02
Change Engine Oil Filter	14.00	Days	08/01/01	WO - 1108-1		08/24/01
Change Hydraulic Fluid	90.00	Days				09/10/01
Change Hydraulic Oil Filter	200.00	Hr	06/19/01	WO - 1094-1	450	
Change oil. 10W30 API, Diesel-CC/CC	7.00	Days	08/01/01	WO - 1108-1		08/17/01
Check Cooling System	1.00	Days				
Check Engine Oil	1.00	Days				
Check hydraulic fluid level	1.00	Days				
Check parking brake, adjust if necessa	14.00	Days	08/01/01	WO - 1108-1		08/24/01
Check rear wheel bearings, and adjust	30.00	Days	08/01/01	WO - 1108-1		09/09/01
Check tires. Inflate front to 10-12 psi, r	1.00	Days				
Clean air cleaner dust cap	1.00	Days				
Clean air cleaner element monthly, or a	30.00	Days	08/01/01	WO - 1108-1		09/09/01
Inspect battery, clean	30.00	Days	08/01/01	WO - 1108-1		09/09/01
Inspect engine belts, and adjust as ne	1.00	Days				
Inspect fuel filter. Diesel - clean, Gas- r	90.00	Days				09/10/01
Inspect hydraulic hoses and fittings	1.00	Days				

Eq. Description: Greens King V (1862G & 1962D) (EQJA51862G)   Current Meter: 450.00

Figure 3-3   A screenshot from SmartShop® software.

ManagerPlus

http://www.managerplus.com/index.php

ManagerPlus software is designed to maintain equipment, vehicles, tools, and facilities. Features include PM, scheduling, tool tracking, repair histories, work order system, and fuel and oil tracking, among others.

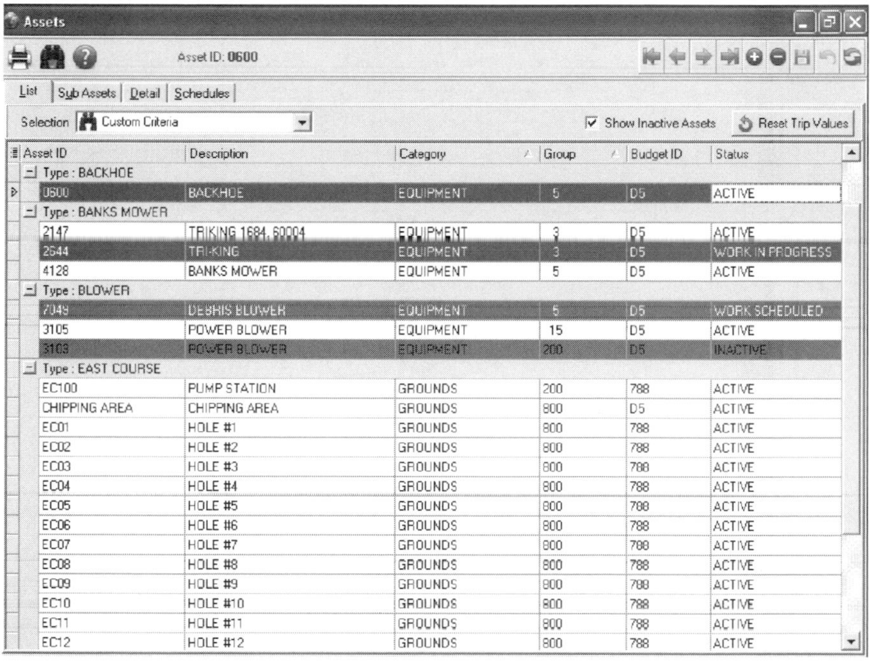

Figure 3-4   A screenshot from ManagerPlus software.

SAMPLE DOCUMENTS

The work order shown in Figure 3-5 was created by the International Golf Course Equipment Managers Association (IGCEMA). It is available on its website at www.igcema.org. Work orders are necessary to track repairs and PM effectively.

Work Order

IGCEMA

Equipment No.:_____

Date: _____

Equipment:_____

Hours:_____

Service____   Defect____

Repair_____   Request____

QTY	DESCRIPTION	Part Number	UNIT PRICE	TOTAL
			Total Cost	

Notes:_____
_____
_____
_____
_____

Time_____

Equipment Technician_____   Date_____

Equipment Manager _____   Date_____

Figure 3-5   A sample work order created by the International Golf Course Equipment Managers Association.

4

Reel Technology

REEL MOWER PRECISION vs. ROTARY MOWER SIMPLICITY

Reel mowers are complex, precision-built mechanisms. These mechanisms are prone to damage, which can result in poor cut quality. If an operator hits a root or the edge of a cart path or even runs over a partially popped-up sprinkler head, damage can occur. In contrast, rotary mowers are very simple and can be constructed to handle a lot of abuse. A basic rotary mower has a quill assembly mounted on a deck. The quill is composed of a blade spindle, a pair of bearings, and a bearing housing. A mower blade with two beveled cutting edges is mounted to the blade spindle. There may also be a simple slip clutch arrangement built on one end of the spindle to protect it from sudden impacts received by the blade. This rugged assembly requires little maintenance other than greasing the bearings and then checking the blade to see if it is running true and sharp. A dull blade can be sharpened on an inexpensive bench grinder, and a bent blade can be economically replaced. Reel mower assemblies require daily checks and adjustments. The reel assembly is composed of a dozen or more parts. Reel cutting units can be damaged beyond repair by moderate impacts and require continuous lubrication, sharpening, and adjustment with expensive, specialized equipment.

When you look at these two methods of cutting grass, a question arises. If rotary mowers are so cheap to operate and simple to maintain, then why do we use reel mowers? This question has two answers.

First, anyone who maintains fine turf quickly realizes that indeed you do use rotary mowers wherever you can. Golf course roughs, areas surrounding sports fields, areas that are not in play, and anywhere you do not need to cut lower than about 1.5" (38 mm) are suitable for modern rotary mowers. The second part of the answer is related to the first. Anywhere you must cut lower than 1.5", you need to use a reel mower to get a quality cut. In many sports, especially golf, there is a demand for ever decreasing cutting heights. Golfers want those extremely fast greens they see in professional tournament play. Those fast greens require extremely low cutting heights that can be achieved only with a reel mower.

Rotary mower engineers are continually improving their mowing decks' performance. Modern rotary decks produce a clean, efficient cut within their cutting height limits. They have the added ability to cut the turf into a fine mulch that can be left behind, with no need to rake or vacuum the clippings. Improvements in reel technology have also continued with tougher frames, simpler adjustments, and reduced maintenance. There may be a time in the future when reel mowers will be supplanted by some new technology, but until then, understanding reels, reel maintenance, and reel sharpening is essential for anyone responsible for maintaining closely mown turf.

GRINDING METHODS

There is nothing in the purview of the turf or equipment manager that is more important or controversial than sharpening reels. There are almost as many techniques and opinions about this subject as there are technicians grinding reels. There are countless influences on grinding methods. Budget, grass type, climate, mowing equipment type, and customer demands have a huge influence on how reels are ground. A facility that never has a cutting height below .250" (6.35 mm) has very different demands on its cutting equipment than a golf course that routinely cuts at .110" (2.79 mm) or less. There are some facilities that grind their reels weekly and some that grind them annually. There are facilities that have several reel grinders and some that have none. Some facilities relief grind every grind, and others relief grind annually or not at all. It is interesting to note that good, consistent cut quality can be achieved with a variety of techniques.

The first of these methods to examine is the relief grind approach. Relief grinding is recommended by most of the major equipment manufacturers. Toro® has a very informative booklet available online at http://www.toro.com/customercare/commercial/education/pdf/98008sl.pdf.

This publication is an excellent overview of Toro's philosophy of reel grinding and covers all the basics of reel maintenance and setup. Toro, John Deere, and Jacobsen recommend relief grinding as well as back lapping their reels with lapping compound. This is another controversial operation that is not universally employed.

Relief grinding entails "thinning" each reel blade by milling (usually done only at the factory) or grinding with a rotating stone in a relief grind–capable reel grinder in the field. Adding relief requires an additional step in the grinding process and, consequently, requires some additional time to sharpen each reel. Relief grinding is promoted as a way to lessen the loading on the reels and the traction unit. Thinner reel blades produce less friction and reduce required horsepower and, consequently, lower fuel consumption.

There are many technicians that flat grind, which is also known as spin grinding. This is a faster operation than relief grinding because setup in the grinder is usually quicker and the relief grinding step is eliminated. Some people prefer to spin grind a reel several times before they relief grind it. The theory here is that a light spin grind can be done several times before the relief is eliminated. Some technicians use spin grinding on a frequent basis, relief grind occasionally, and skip back lapping altogether.

Bed knife grinding is the other half of the equation necessary to produce a decent cut. Bed knife grinding requires a second grinder in your inventory. In order to get a precise cut across the width of each reel blade, the bed knife must be ground to the correct angle and parallel to the reel blades. There are two areas of concern: the front face angle and the bed knife angle. Both of these angles are specified in the equipment manufacturer's service or owner's manual. Some of the latest bed knife grinders have the correct angles for most equipment stored in the microprocessor memory. This can be a time-saver.

BACK LAPPING

As previously mentioned, back lapping remains controversial. There are many technicians who do not back lap. Back lapping is employed to hone the reel blades and, according to John Deere®, to remove the burrs left from grinding. Most equipment manufacturers caution against using oil or grease-based back lapping compound because they are hard to remove from the reel. If you use an oily compound, the temptation is to attack the stuck-on residue with high-pressure water. This can be fatal because water at high pressures can force the compound and water into the bearing areas, which rusts and destroys the bearings. Most equipment manufacturers are reluctant to recommend cleaning with water, especially high-pressure water. Air pressure and mechanical means of removing grass and debris are preferred because of the inherently corrosive effect of water. Some courses use reclaimed water to wash equipment, and the concentration of herbicides, pesticides, and fertilizers in the water can corrode equipment very quickly.

GRINDING EQUIPMENT

There are a number of different manufacturers of excellent grinding equipment. Foley United has been building grinding equipment since 1934. The Simplex Ideal Peerless reel grinder company traces its origins to 1902. Bernhard and Company, which began more than 50 years ago, manufactures the Express Dual and Anglemaster brands of grinders. There are several other brands of grinders available in the marketplace, but these are the major players.

Most grinder manufacturers have both spin grinding and relief grinding capabilities in their product lines. Once

Figure 4-1   The Express Dual grinder (foreground) and Anglemaster grinder (background), both shown here at the Old Collier Golf Club in Naples, Florida, are manufactured by Bernhard and Company.

you decide what capability you want, you can obtain bids, arrange for demonstrations (or even loaner equipment), and make your purchases. Be sure to price replacement stones and other wear items. These hidden costs can have a big impact on your budget.

Frequently, a decision about which grinding equipment to purchase is a simple matter of cost. Lower budgets dictate lower-priced machines, even refurbished or used equipment. Grinding equipment has become computerized and automated. These grinders "remember" how you ground each reel the previous time and can utilize this information on the new grind. To effect this amazing feat, a microprocessor and memory chip are employed inside the grinder control panel. The newest grinders are also enclosed or screened to reduce

Figure 4-2 This ACCU-Pro grinder comes from Foley United.

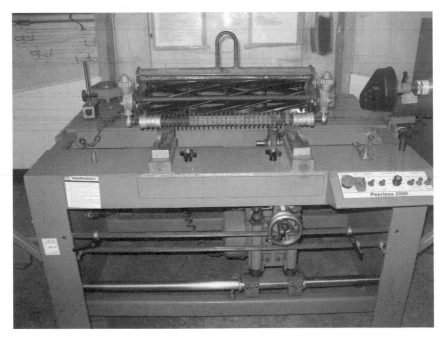

Figure 4-3 This Peerless 2000 grinder comes from Simplex Ideal Peerless.

noise and prevent sparks, metal, and grinding particles from spreading around the shop. Many high-budget courses trade grinders every few years. If you do not have a large budget, keep in touch with a facility that does. If you can buy a good used bed knife or reel grinder at a trade-in price or a few dollars above trade-in, then you and the seller will both benefit. Most grinding equipment is sold by a local turf equipment distributor. If you have a distributor who gives you good service, brand selection is simplified. Consider simply buying whatever brand they are selling.

Admittedly, all this information has conflicting points. Do I back lap or not? Do I spin grind or relief grind or both? All the preceding information is an oversimplification of a complex subject that has kept scores of engineers, technicians, and turf managers engaged for many years. There are no easy answers, but remember that you will find your own way if you experiment. You will find a way that works for you, your turf, your local conditions, your budget, and your customers.

REELS

Reel mowers have been evolving for more than 100 years. Minor and sometimes major innovations occur continuously. In recent years, frames have been strengthened to keep the reel, bed knife, and roller(s) in parallel. Bearings are tougher and better sealed. Metals have been improved in the bed knives and reels. Height of cut (HOC) adjusters have been simplified and made more secure. All these improvements have been beneficial to cut quality, but even with all these efforts from manufacturers, there has been a critical element that is sometimes overlooked by some technicians. Reel mower components do not stay parallel. The bumps, bangs, and scrapes of everyday use knock the reels akimbo. This results in misalignment of the rollers, bed knives, and reels.

Reels will not cut correctly when their major components—the reel, bed knife, and rollers—are misaligned. Many technicians have fussed over reel to bed knife clearances, HOC adjustments, and other tweaks but failed to realize the root of the problem: the three major components (reel, rollers, and bed knife) were not parallel. The technicians who understand this phenomenon have devised precise ways to check and adjust for this problem. The fabrication of a precision granite surface plate covered with a protective steel sheet makes a perfect tool for checking parallelism. The roller(s) can be precisely paralleled to the bed knife and reel blades on this plate. This essential first step guarantees that subsequent adjustments, such as reel to bed knife clearance and HOC, are effectively accomplished.

Figure 4-4   This leveling table at the Golf Club of the Everglades in Naples, Florida, was fabricated by Jason Ticknor.

ROTARIES

The increasing popularity of rotary mowers has become an interesting trend in the turf industry. Consumer mulching mowers were marketed with more enthusiasm than engineering when they were introduced in the 1970s. The inability of these units to perform as promoted and the manufacturers' focus on homeowner units resulted in very little notice in the fine turf community. The original designs were predominantly in the walk mower category, which also reduced their appeal. The latest designs, such as the Toro® Contour™ Plus, Jacobsen Envirodecks™, and John Deere's independent deck design have become popular for rough and surround mowing in recent years. These small (21"–27") decks, ganged in five- or seven-unit arrays, follow ground contours effectively and mulch the clippings very finely. When it is necessary, most of these decks can also be converted from mulching to discharge chute operation in just a few minutes. The combination of convertibility, effective cut, fast cutting speeds (up to 7.5 mph), and reduced maintenance has driven the appeal of these designs.

Rotary mowers are not effective for cutting heights much below 1.5" with most grasses. The current designs also do a creditable job up to about a 4" HOC. The trends in average rough height vary from region to region and with course or field types, but this range (1"–4.75") of rotary rough HOC is sufficient for most turf operations. This versatility and the generally lower costs of maintaining rotary mowers have made this type of mower increasingly popular.

ADJUSTING FOR HEIGHT OF CUT

Many people have only a partial understanding of HOC. You might hear someone say, "Oh, we are cutting at .090"," or "We always cut at less than .105"." A lack of specificity in these statements often means a lack of understanding about HOC. Are they referring to the bench set measurement before the unit left the shop? Are they referring to the actual height of the grass measured with a prism or other such turf height measuring device? What is the HOC with the grass catcher full? Empty? *Height of cut* can be a nebulous term when misunderstood and/or poorly defined. If you are sending a number of mowers out to cut the turf, the most important thing is consistency. Each tee mower, greens mower, or fairway unit in a group must be set to the same effective HOC. If you fail to establish a uniform way of measuring HOC, you will fail the consistency test and your customers will complain.

It may seem obvious, but in order to reach this consistency of HOC, all mowers for a particular task must be set with the same tools. An accurate dial indicator and bar specifically designed to adjust HOC, such as the ACCU-GAGE® from Accuproducts, is recommended.

To verify the actual height of cut, another useful tool from Accuproducts is the PRISM-GAGE™. This is a graduated prism that is placed on the turf; the reflected height of cut appears on the graduated scale. When you need a very accurate measurement of actual HOC, this is a good tool.

Figure 4-5   This PRISM-GAGE™ comes from Accuproducts.

TIRE PRESSURE

Tire pressure is mentioned here because it can have an effect on cutting quality. Overinflated tires can result in continuously bouncing traction units. Sometimes, even the best reel or rotary deck suspension designs cannot stay on the ground when this happens. The effect is an undulating cut that can be misdiagnosed with frustrating effects. Other effects of excessive tire pressure are increased compaction, lack of traction, tire spinning, increased fuel consumption, and turf damage. All traction units are designed to obtain maximum traction and fuel efficiency with a specific tire pressure.

Low tire pressure can have equally damaging effects. Reducing tire pressure below the recommended amount can result in overheated tires and accelerated tire wear. If tire pressure is too low, the tire can slip on the rim. Tube-type tires that slip will rip the valve stem out of the rim, resulting in a flat tire. If tire pressure is too low, the tires will lose their grip and slip. This is very dangerous when the operator is on a side hill.

ROTARY BLADES/WEAR/SHARPENING

Rotary blades are easy to maintain if a few basic precautions are observed. Rotary blades are subject to primarily three potential malfunctions: becoming bent, dull, and eroded. Bent blades produce an uneven cut. Sometimes, there will be arcs and swirl patterns in the turf that indicate a bent blade. Dull blades damage the turf by bruising the tops of the grass, leaving a ragged cut and possibly promoting disease. If a rotary blade becomes eroded, the lift surface on the back side of the cutting edge will get thin and begin to curl or separate from the blade. This is a potentially dangerous situation because the curl of metal that was the lift surface can be thrown from under the deck at high speed. Eroded lift surfaces reduce the lifting action of the blade, and the grass that is not lifted will remain uncut. This will leave shaggy, uncut patches in the turf.

Another problem to guard against is cutting when the grass is very wet. Cutting when the grass is wet will produce clumps and prevent a mulching mower from recutting the grass into fine clippings. This defeats the mulching process; instead of the clippings returning to the base of the grass stems as nutrients, they remain on top of the turf, causing yellow and brown patches.

Three steps need to be taken when a rotary mower blade is sharpened. First, inspect the blade for wear and check the track. A Magna-Matic® blade balancer is equipped with an adjustable track rod that can be used to determine whether a blade is bent. To use the track rod, first, adjust it so the ball end touches the blade tip at the cutting edge. Next, rotate the blade 180 degrees. If the track rod ball end touches the blade, the blade is straight. Bent blades cannot be straightened! Replace them. Eroded lift surfaces also mean the blade must be replaced.

Next, sharpen the blade with a bench grinder or dedicated rotary blade sharpener (see the following illustration). There are a number of manufacturers building automatic and semiautomatic rotary blade sharpeners. Foley United, Bernhard, and Magna-Matic manufacture rotary blade sharpeners. Some of these units feature automatic feed and coolant systems. Follow the original contour of the cutting edge. Keep the ground area smooth and blended into the transition area where the blade is unsharpened. Notches and gouges in the blade encourage erosion and accelerate wear.

Finally, balance the blade. This is done with the Magna-Matic® balancer. If the blade is heavy on one end, it will rotate downward. Remove more metal from the heavy end until the blade stays parallel to the ground.

Figure 4-6   This example of a blade balancer is the Magna-Matic® blade balancer with track rod.

Figure 4-7   This example of a blade sharpener is the Bernhard Rota-Master rotary blade sharpener.

Be sure to caution your operators about the signs of bearing failure and damaged or bent rotary mower blades. If they hear unusual sounds from the mower deck, perceive more vibration than normal, or observe cutting defects, they need to shut down the equipment and check the mower deck spindles, blades, and drive system. An alert operator is always your best defense against catastrophic failure and unnecessary downtime.

5

Selecting Tools and Equipment for the Shop

The Right Equipment for Your Shop
Must-Have Tools
Special Tools

THE RIGHT EQUIPMENT
FOR YOUR SHOP

Shop equipment can be either a money and time-saver or an endless economic drag on your budget. You and/or your equipment manager must create a realistic plan for choosing, stocking, and maintaining the right tools.

Before you ask yourself what equipment is necessary for your shop, you need to answer a few basic questions. What is your budget? What tools and equipment are already in your inventory? What are the essential tools for making your shop functional? In order to more easily grasp the scope of tooling your shop, it might be a good idea to initially restrict your list to those items that apply directly to repair and maintenance. This has been done in the following list. The list is alphabetical, and the zones are not totally inclusive. Every shop will have a slightly different set of zones. Every shop will also have some variations in tooling. Some equipment may appear in one zone in your shop and in a different zone in a shop down the road. Use this list as a guide for stocking your new shop or aligning the inventory of your existing shop. Items that require additional guidelines or comments will come from this list. Arguably, the most important item on this list is the equipment lift found in the lifting zone.

MUST-HAVE TOOLS

The Master List of Must-Have Tools shown in Figure 5-1 includes the basic tools you need for an efficient shop. This list has been refined by a number of turf equipment technicians over the past few years. Tools and techniques change, turf equipment changes, and the demands of the customer change. Five years ago, if you suggested to a golf course superintendent that roughs would be cut with a rotary mower, you would have been laughed off the course. Today, rotary mowers are used extensively for the deeper roughs players are asking for. This is a complete reversal of the previous trend both in mowing equipment and course conditions. Rotary mowers require a heavy-duty bench grinder or a specialized automated blade grinder. You need an accurate blade balancing and track measuring device. The introduction of rotary mowers was one of those incremental changes that reshaped shop tool requirements.

If you look through the alphabetical list, you will notice tools designated with a ✿. These are the optional tools that will increase shop production. In some instances, such as the granite surface plate compared to the steel plate, reel setup becomes more accurate. Occasionally, there will be tools listed that are found in only a few shops. These tools might be categorized as the ones you would be "nuts to have" for most facilities. Some maintenance facility superintendents and technicians swear by these tools and consider them essential. Sometimes, these tools are regional in nature. The specialized

Optional	Tool	Notes
	2.5 ton bottle jack	2 minimum
	3.5 ton floor jacks	2 minimum
	4 drum containment pad	
	4" vise	
	6 drum containment pad	
	6" hand grinder	
☼	6" vise	Wilton
	8" step ladder	
☼	8" bench grinder	
%	8" lathe	
	10 gal oil can w/casters	
☼	10" cold metal saw	
	12 ton hydraulic press w/blocks	
	Air chuck	2 ball foot and extended fill
☼	Anvil 55#	Northern Tool + Equipment
	Battery carrier	
☼	Battery charger	Battery Tender®
	Battery charger	200 amp w/wheels
	Battery/charging system tester	
	Battery cleaning tool	
	Battery pliers	
☼	Battery powered grease gun	
	Bearing installer set	Special tool
	Bed knife facer	R&R
	Bed knife grinder	Angle Master 3000
☼	Belt sizing gauge	Goodyear or Dayco 93860
	Belt tension gauge	Krikit I belt gauge
	Drill index	Special tool
	Drill press	
	Drum dolly	
	Drum transfer pump	
	Equipment lift	
	Face shields	2 minimum
	Fastener cabinets and supplies	
	Filter crusher	
	Flammable storage cabinet 20 gallon	

Figure 5-1   Master List of Must-Have Tools. ☼ represents optional tools.
% represents tools found only in a few shops.

(*continues*)

Optional	Tool	Notes
☼	Flammable storage cabinet 45 gallon	
	Fluid evacuator	Mity Vac
	Gear puller set	Special tool
	Hand-powered grease guns	3 to 5 guns depending on types of lube required
☼	Hand truck or dolly	
	Heat gun	
	Heavy duty workstation/bench	Strong Hold
	Height of cut gauge	2 or more Accu-Gage or Foley United
	High lift jack stands	1 or 2
☼	Horizontal mill	
☼	Hydraulic press 12 ton	
☼	Hydraulic work station	Hydraulic hose creation
	Jack stands	
	Jump box	Jump-N-Carry 660
☼	Layout table	
☼	Lester tester	Lester Electrical Corp.
	Mechanics floor creeper	
	Metal storage cabinets	As needed for special tools
☼	Metal storage rack	
%	Mill	
☼	Mobile lift table	Trion DL 1300
☼	Motorized band saw	
	Neoprene apron	
	Neoprene gloves	2 pairs
	Nitrile gloves	Box of 100
	Oil filler 6 qt	
	OSHA approved blow gun	
☼	Oxy/acetylene rig	
☼	Pallet jack	Jet Brand
☼	Parts "blow off box"	
	Parts washer	
☼	Plasma cutter	
	Pneumatic blow gun	
	Portable air tank	
	Protective goggles	2 pairs minimum

Figure 5-1   (continued) Master List of Must-Have Tools. ☼ represents optional tools. % represents tools found only in a few shops.

Optional	Tool	Notes
	Reel adjustment bench plate	Metal or stainless steel over granite
	Reel grinder	Express Dual 3000
	Rolling tool box	
	Rotary blade balancer	Magna-Matic
☼	Rotary blade sharpener	Foley
	Shop vacuum	
	Smaw welder	
	Tap & die set	Special tool
	Terminal spreader	
%	TIG welder	
☼	Tire cage	For on road tire service
	Tire changer	Heftee or Coats
	Tire gauges	Accutire, Accugauge, Monkey Grip
	Tire test tank	
☼	Tow ropes	
%	Turf evaluator	Special tool
	Wash tank	
	Welding pliers	
	Wheel chocks	2 minimum

Figure 5-1   (continued) Master List of Must-Have Tools. ☼ represents optional tools. % represents tools found only in a few shops.

equipment necessary to break ice on putting greens is one example. These tools are indicated by a question mark.

The Lubrication Zone

The lubrication zone should be located adjacent to the lifting zone and contains all the necessary equipment for greasing, lubing, and changing fluids. This equipment includes the following:

- Handling system for filling and evacuating liquids (simple or more complex and expensive oil removal equipment)
- Grease gun (battery powered or hand operated)

This zone requires a handling system for filling and evacuating fluids. Choose the system that your budget allows.
A simple, inexpensive system consists of several extendable funnel-equipped catch cans. One is dedicated to oil and
the other to such fluids as antifreeze or fuel. These portable oil drain cans have a capacity ranging from 8 to 20 gallons, have casters on the base, are equipped with extendable
funnels, and cost from $60 to $130 each (www.gesusa.com).
If you have to lift and pour the waste oil out of these containers, remember that a gallon of oil weighs approximately
7.5 pounds. Obviously, a 20-gallon container will be tough
to lift and empty. If you have an oil suction removal system,
this is not an issue. Oil suction removal systems can be very
expensive, especially if they have a large remote tank and
several suction points; it is not unusual to spend more than
$5,000 for a sophisticated system.

One particularly impressive suction system has the used oil
storage tank in a room on the end of the building and a remote
removal port mounted on the outside wall. This system keeps
the oil recycler service's oil-soaked hose out of your shop. If
you have looked closely at one of these hoses, you will understand why this is a definite plus. A less expensive used oil
suction system can be constructed with a tank on a cart. The
cart is also equipped with a battery-powered, reversible pump
used to empty oil catch pans and cans. The cart can be rolled
to the main used oil storage tank and the pumped reversed to
transfer the oil. This system requires more oil handling than a
dedicated suction system but is much more economical. Also,
a hand suction gun should be included for those tasks for which
the evacuation system will not fit or is too hard to control.

An additional item in the lubrication zone and the primary
tool in this zone is the grease gun. Grease guns are one of those
boring and ubiquitous shop tools no one thinks about. If you
choose the wrong grease gun, however, you can do more damage

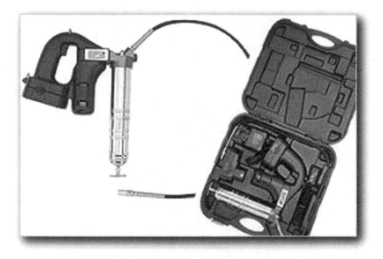

Figure 5-2   This battery-powered grease gun is from Plews & Edelmann.

Figure 5-3   The port is the outside suction and fill spot for the Old Collier Golf Club oil-handling system. Waste oil is extracted and fresh oil is delivered through this port on the outside wall of the maintenance facility.

Figure 5-4   This is the room on the other side of the wall from the port shown in Figure 5-3. The silver checker plate box in the center of the photo is the back of the port.

Caution

OSHA requires that all used oil containers be labeled "Used Oil." See Chapter 9 for additional information about fluid handling, waste material record keeping, oil filter handling, spill kits, used antifreeze, and gasoline and diesel fuel disposal. If local or state laws demand it, you may have a filter crusher in the lubrication zone.

When storing used oil and hydraulic fluid, remember to follow the required containment procedures outlined by the state authority. Usually, the minimum requirement is a spill containment pallet for each drum. You may also be required to provide containment for new oil and hydraulic fluid drums.

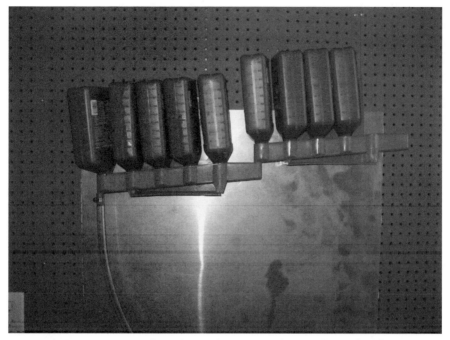

Figure 5-5   Here is an excellent device for getting the last drop of oil from quart cans.

than you might realize. The grease seals on reels and other equipment can be dislodged or distorted by improper greasing. Damaged seals allow dirt to enter the bearing areas and destroy them. Notice that there are two types of grease guns shown on the Master List of Must-Have Tools in Figure 5-1. Battery-powered grease guns are a relatively new innovation but are controversial because they usually do not have adjustable pressure regulation and are capable of pressures in excess of 6,000 pounds. Some hand-operated grease guns are capable of pressures up to 10,000 pounds per square inch (psi). This much pressure can force a seal out of its housing. Hand grease guns with adjustable pressures are available. The most desirable grease guns have pressure limit valves. Some are limited to 3,000 psi maximum pressure and are much easier on seals.

Air line–powered grease guns are not recommended without careful air line pressure regulation. Most air line–powered grease gun manufacturers recommend no more than 30–40 pounds per square inch of air line pressure to protect delicate seals. These guns can invite abuse from an improperly trained technician using excessive air line pressure.

The Lifting Zone

The heart of the lifting zone is the equipment lift, but this area also includes the following:

- Lift Attachments
- Jack or Equipment Support Stands
- Bottle Jacks

Figure 5-6   This lift is located at the Ritz-Carlton® Members Club in Sarasota, Florida. The oil and air supply reels attached to the lift are handy additions.

- Floor Jacks
- Hand Truck
- Lift Stands
- Mobile Lift Table
- Pallet Jack

The equipment lift is the most important tool in the shop. If you do not have a lift, find a way to put one in your budget and make it a top priority. An equipment lift is a time-saver and a back saver and will improve the quality of unit maintenance and repair. Good technicians you interview will make initial judgments about shop quality by the existence or absence of a lift. You can make a similar judgment about the quality of your interviewee.

First, ask yourself the following questions: What do we need to lift? How much lift can we afford? If you do not need to lift pickup trucks or other on-road vehicles, lift cost is less because lifting large road vehicles usually necessitates a second lift. If you have fleet management responsibility for pickups and autos, budget for a second lift, since most golf equipment lifts are not well suited for both tasks.

Equipment lifts come in several configurations. Look for the lifts designed specifically for turf equipment. Some brands may simply be modified automotive lifts. Look closely at any attachments you think you need. Attachments are available to lift walk-behind mowers to a comfortable working height, adapt to small-wheeled units like aerators, and extend capacity lift arms for utility vehicles with sun tops. Check to see whether the accessories are easy to install and remove. A lift with all attachments can be quite expensive. A modified automotive lift (a poor choice) could be a third of the price of a fully equipped turf equipment lift and not be a bargain. There are many golf lift brand choices. The best approach to lift selection is to first determine the amount

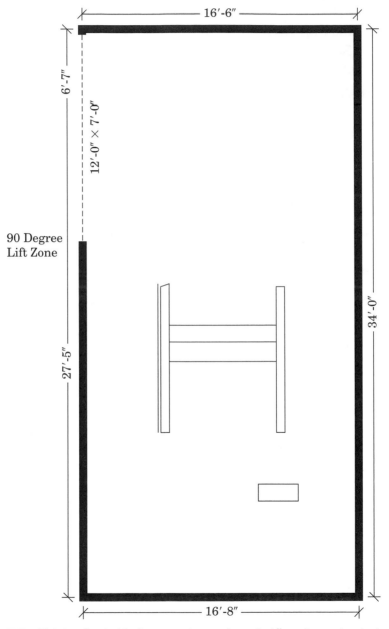

Figure 5-7   This is a basic 90-degree entrance layout. All equipment entering the lift has to turn 90 degrees.

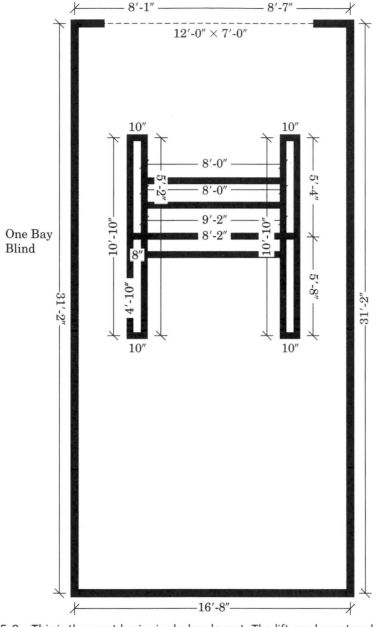

Figure 5-8 This is the most basic single-bay layout. The lift can be entered without turning, a common design.

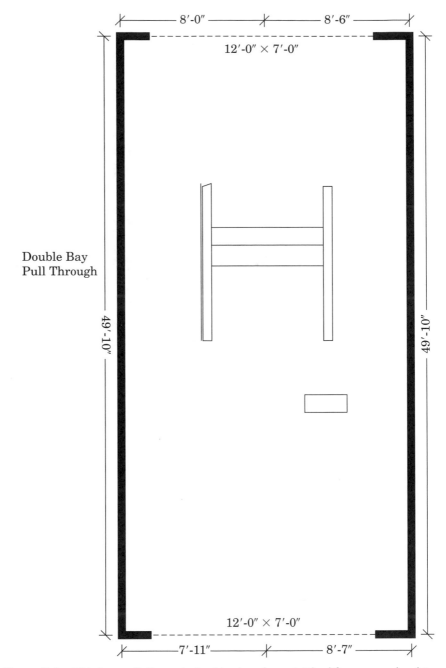

Figure 5-9   This is a pull-through double-door layout. The lift area can be driven through. This is a convenient layout, but it sacrifices space.

you have to spend. Next, survey the lifts in use in your area. Ask a lot of questions:

- How do the attachments work?
- Have there been any service or safety issues?
- Are the lifts actually being used (As strange as this question may seem, there are lifts that are so complicated to set up and difficult to use that they become giant storage racks for tools and grease guns. Even a well-designed lift that is poorly located on the shop floor may end up unused. Location is as critical as the right brand and model.)

Next, compile a list of popular brands and their respective prices. Some of the available brands include Trion®, Golf Lift®, Manitowoc, Heftee, Mohawk, PMW, Ben Pearson,

Figure 5-10   This Golf Lift® brand lift is located at the Old Collier Golf Club in Naples, Florida.

John Bean, Rotary Lift®, and Snap-on. You should be certain to include delivery and installation costs for each lift you price. You may discover that some brands are automatically quoted with installation costs included.

The last consideration with a turf equipment lift is capacity. How many pounds will it lift? This is the least important specification for turf lifts and, regretfully, the first one that many buyers seek. All top-quality turf equipment lifts have sufficient capacity to lift the heaviest fairway units. If you anticipate lifting utility tractors or ancient tractor-style fairway units, you are beyond the capacity of a turf equipment lift and need an automotive-type lift. The most important

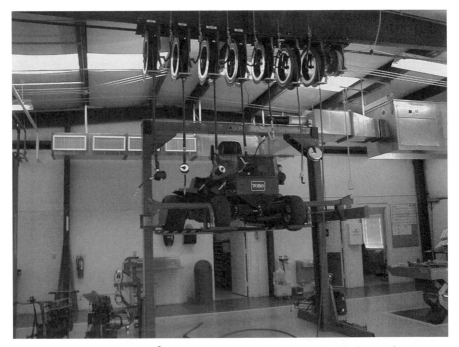

Figure 5-11   This is a Trion® brand lift at Cherokee Country Club in Atlanta, Georgia. Note that it has a cross bar at the top for stability. The hose reels in the foreground are capable of supplying any necessary fluids, electricity, and compressed air required at the lift.

Figure 5-12   This is a PMW lift located in the Tournament Players Club of Tampa Bay (Florida) maintenance facility. The mezzanine storage area is visible in the upper right corner of the photograph.

features for turf equipment lifts are speed of setup and ease of use. Choose your lift wisely, and locate it carefully.

Additional tools for this zone include jack or equipment support stands (four or more), two 2.5-ton bottle jacks, one or two 3.5-ton capacity floor jacks, a hand truck, two high lift stands, and a mobile lift table like the Trion® DL 1300. The Trion® is a very safe way to handle heavy reels and other cumbersome equipment that needs to be raised off the floor. Similar products are available from Golf Lift®, Southworth Products Corp, and others. A case can be made for including a pallet jack in your inventory if you receive fertilizer, seed, and other palletized materials.

The Grinding Zone

- Reel Grinder
- Bed Knife Grinder
- Rotary Blade Grinder
- Bench Grinder
- Rotary Blade Balancer
- Front Facer
- Height of Cut Gauges
- Leveling Table
- Rotary Blade Balancer

The grinding zone is another critical zone, and the most expensive equipment in the shop is concentrated here. The reel grinder, bed knife grinder, rotary mower blade grinder,

Figure 5-13   This is the grinding zone at the Old Collier Golf Club in Naples, Florida. Note that the fabrication zone is adjacent to the grinding zone. These two hazardous, spark-producing zones are well separated from the rest of the shop.

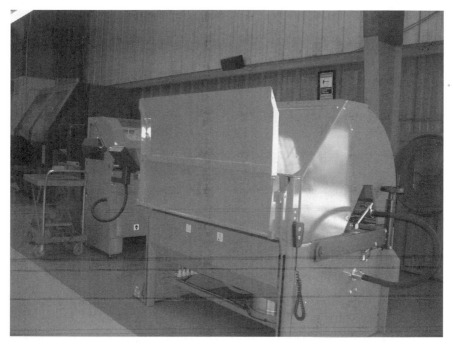

Figure 5-14   This is the grinding zone at the Black Diamond Ranch in Lecanto, Florida. Their equipment manager uses two different brands, Foley and Bernhard, for reel grinding, an approach used in a number of facilities.

and bench grinder are the basic tools in this area. Some additional tools kept in this zone include a front facer, rotary blade balancer (such as a Magna-Matic®), height of cut gauges from Accu-Gauge or Foley United, a leveling table, back lap machine, reel storage racks, and the hand tools used in reel setup. The hand tools for this zone may be supplied from the mobile toolbox or are sometimes a dedicated set kept permanently in this zone.

The Tire Repair Zone

- Tire Cage (for on-road tire service)
- Tire Changer
- Air Chuck

- Blowgun
- Tire Test Tank
- Wheel Chocks
- Tire Gauges
- Portable Air Tank
- Tubeless Tire Repair Kit

If you must service on-road tires, you will have an additional set of tools and equipment that is not necessary for turf tire service. On-road service requires a tire safety cage and a larger capacity tire changer. All tire zones need a tire repair station, which includes an air chuck, a blowgun, a water test tank, wheel chocks, at least two accurate tire gauges (including a low-pressure model), a portable air tank, a tubeless tire repair kit with a needle, a rasp, string patches, and cement. Tire gauges are notoriously inaccurate, and tire pressures on turf equipment are critical to quality of cut and traction. Pencil-type gauges tend to become less accurate with age. A gauge two or three years old can be off by four or five psi. The newer digital gauges hold promise. They have proven to be more accurate, and some have a self-calibrating procedure that needs to be exercised regularly. Digital gauges, like their mechanical counterparts, can be damaged if dropped. Some of the Accutire® (model MS-4000) digital tire pressure gauges are accurate but very hard to use because it is difficult to get the snout of the gauge squarely on the valve. The later Accutire® (model MS-5510 B) digital design with the short hose is easier to use. The Accu-Gage analog gauge is similar to the Accutire® digital, with a short extension hose that places the gauge away from the valve for easier viewing; it also solves the problem of aligning the hose end and valve. Good technicians keep an accurate gauge stored securely in their toolbox and use it to periodically check the accuracy of the tire zone gauges. Discard any gauge that is off by more than one or two psi.

Figure 5-15   This tire repair zone is located at Lake City (Florida) Community College. The shadow board technique is a good way to provide a quick inventory of your tools.

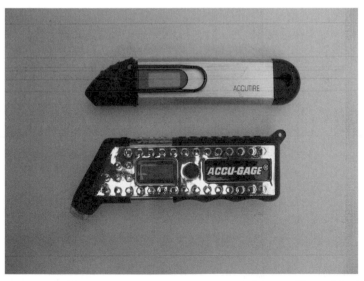

Figure 5-16   A reliable, accurate tire gauge is essential. You will need both low-pressure (0–30 psi) and medium-pressure (0–100 psi) gauges.

The Cleaning Zone—Interior

- Wash Tank
- Gloves
- Goggles
- Blow-Off Box with Fan

This zone's primary tool is the wash tank or cabinet. There are numerous options for small parts cleaning. The current trend, the result of environmental regulations, is a heated, aqueous solution system. Water-based systems do a decent cleaning job and avoid the hazards associated with solvent-based cleaners. Solvent-based systems are still available. Other options include enclosed agitation systems that use heated, high-pressure water and a special soap in a closable cabinet. These giant dishwasher-like cabinet systems are best suited to batch cleaning because the heating element must run for several hours to bring the solution up to temperature; subsequently, the operating cost is almost as much to clean a single part as a full cabinet. These units come in a variety of sizes, from a 2.2-gallon tabletop unit to walk-in industrial sizes. Hydro-Blast™, Graymills, and Landa® are popular brands. Currently, the most prevalent systems continue to be solvent based and available from a servicing supplier like Safety-Kleen. These companies will set up a service schedule and exchange your dirty solvent for filtered, recycled solvent on a periodic basis that suits your needs. Safety-Kleen also now offers aqueous systems. Be sure to keep nitrile or neoprene gloves, a neoprene-coated apron, and splash goggles near the wash tank.

The cleaning zone also requires a compressed air supply to blow off freshly washed parts. Compressed air may also be necessary to power some of the agitating cleaning systems. A nice addition to the cleaning area is a five-sided box that is vented outside the building. This blow-off box keeps the air, solvent, or water and remaining debris contained rather than splattering it on the shop walls. A sheet metal shop or commercial HVAC company can build a galvanized box with a duct for you. Mount the box on a workbench, route the duct through the shop wall, and install a switched, explosion-proof duct fan to pull out the debris.

The Repair Zone

- Rolling Toolbox
- Storage Cabinet/Workbench
- Hydraulic Press with Plates

The repair zone is normally organized near the lift. A rolling toolbox and a mobile workbench will keep this zone flexible. If your budget allows, the most durable workbenches are available from Knaack and Strong Hold. Knaack has the "War Wagon" model with six compartments, and for ultimate storage capacity, it stocks a bench that has a 3,400 pound load rating. Strong Hold (www.strong-hold.com) has a cabinet workstation with an upper storage compartment that is beautifully built. If you visit the company's website, be sure to look at all the workstations available. Strong Hold has some good solutions for mobile storage and a nifty keyless locking system for its workbenches. Be sure that any workbench you purchase can be fitted with polyurethane casters. These wheels make the heaviest bench easy to push and maneuver. It sounds obvious, but larger wheels roll more easily and better handle encounters with floor debris.

A 12-ton hydraulic floor press with press plates is handy for pressing and removing antifriction bearings, bushings, and shafts.

Figure 5-17   This heavy-duty Strong Hold® brand workbench can be ordered with or without caster wheels. Polyurethane wheels are recommended.

The Flammable Storage Zone

• Flammable Storage Cabinet(s)

This area contains a yellow storage cabinet specifically designed to house flammable chemicals, oils, and greases. This is sometimes erroneously referred to as a "fire cabinet."

Obviously, fire or sparks in or near this cabinet is to be avoided. Be sure to check with your local authorities for specific recommendations before investing in a flammable storage cabinet.

The Battery Zone

• Terminal Tools
• Pliers

Figure 5-18   Flammable storage cabinets are essential for safe storage of paints and other flammables.

- Baking Soda
- Splash Shield
- Goggles
- Neoprene Apron
- Battery Filler
- Jump Box
- Battery Tender®
- 200-Amp Battery Charger
- Load Tester
- Battery Acid (if required)
- Battery Bench

The battery zone tools include a terminal cleaning tool; battery pliers; a terminal spreader tool; a battery carrier; a large box of baking soda to neutralize battery acid; a splash shield; goggles; a neoprene apron; neoprene gloves; a battery filler; a jump box/power supply, such as a Jump-N-Carry 660®; a small battery charger, such as a Battery Tender®; a large roll-around 200-amp jump-start capacity charger; and a battery load tester (a Lester tester from Lester Electrical is the best choice if you must test battery-powered utilities, golf cars, or other 36/48-volt equipment). A number of options exist for battery and charging system testing; refer to the list for these. You may need a supply of electrolyte if you service dry charged batteries. You might forgo the roll-around charger if the jump box and small battery charger turn out to best suit your needs.

Battery workbenches need to be noncorrosive. You can build an inexpensive 2' × 4' wooden bench from a standard workbench kit

Figure 5-19   This battery station at the Golf Club of the Everglades in Naples, Florida, is compact and well organized. A fire extinguisher in this area is recommended.

found at most building supply stores. A plastic workbench is another option but is more expensive than wood. Plastic is noncorrosive in contact with battery acid and has the added bonus that it can be rinsed with a baking soda and water solution to neutralize any spilled acid. Wood benches will have to be replaced periodically as they become acid soaked unless you cover them with a noncorroding material. A thermoplastic countertop is one option. These are available from several sources. One technician found a six-foot-long plastic countertop that was damaged on one end. He bought it for $10, salvaged four feet of the countertop, and covered his workbench with it.

The Fabrication Zone

- Oxy-Acetylene Torch Kit
- Shielded Metal Arc Welder (SMAW)
- Drill Press
- Layout Table
- Metal Inert Gas (MIG) Welder (substitute for SMAW)

This area can be as elaborate as your skills, budget, and local demand dictate. The expectations of management and the services provided by your predecessors may dictate your fabrication capabilities. There is a long tradition of innovation and invention in the turf equipment industry. Turf equipment technicians are frequently talented inventors. Every technician is certain there is a way to improve a piece of equipment. There is even a United States Golf Association Green Section regional representative who collects slides and photos of these inventions to share with turf equipment technicians in his region. Often, turf equipment salespeople have photos and sketches of these innovations. There are many rewards for supporting appropriate innovation.

Figure 5-20   The welding zone at the Ritz-Carlton® Members Club in Sarasota, Florida, has a unique "spark catcher" grate on the floor under the welding table.

All turf equipment shops need to be able to heat and weld metal. An oxy-acetylene torch, a small shielded metal arc welder or "stick" welder, a drill press, a small layout table, a small anvil, a vise, a cold saw, and a metal storage rack are the basic items required in a modest fabrication zone. The advent of inexpensive MIG welders has turned almost every technician into an amateur welder. These welders are sometimes referred to as "wire welders" because they use a gunlike handle fed by a spool of welding wire either coated or enveloped in a gas like CO_2 or argon. They are easy to use and require minimal skill. MIG welders will be the welder of choice in most shops. You can add a plasma cutter, large fabrication table, shear, motorized band saw, tungsten inert gas (TIG) welder, and even a mill, lathe, and other

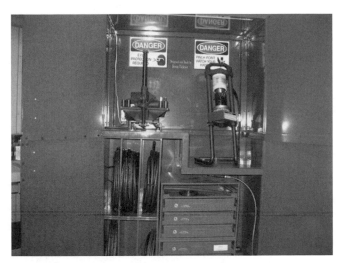

Figure 5-21   This hydraulic hose fabrication cabinet was designed and built by Jason Ticknor at the Golf Club of the Everglades in Naples, Florida. This cabinet keeps all the necessary equipment and supplies clean and organized. This is a nice example of what can be created by an innovative equipment manager in a modest fabrication zone.

(Photo from Jason Ticknor)

accessories to create a complete machining and fabrication shop. This additional expensive, elaborate equipment is usually not necessary.

The Hydraulics Zone

• Hydraulic Hose Fabrication Kit

This zone contains the tools, hoses, and fittings necessary to create custom hydraulic hoses. This approach to hose replacement has a proven payback. Creating your own hoses yields reduced downtime and a guaranteed fit. Hose suppliers will usually offer a package deal if you buy the hose cutting and crimping tools with the hose and fittings. The vendor supplies the hoses and fittings at a lower price if you buy its system. You can either dedicate a workbench to

Figure 5-22 Here is a creative approach to organizing the hydraulics zone. Stephen Tucker, the equipment manager at the Ritz-Carlton® Members Club in Sarasota, Florida, came up with this solution.

accommodate the hydraulic tools and supplies or build your own workstation. There are several ways to construct a hydraulic workstation. See the photo above for some ideas.

SPECIAL TOOLS

Special tools are an expensive and necessary aggravation. Because they are expensive, you will need to establish a policy to determine what constitutes a special tool and who should pay for such tools. In some organizations, all tools are owned by the company. This is an increasingly popular option because the tools necessary for daily setup and service remain with the shop if the technician leaves. Other options to this procedure include the technicians buying all tools (usually with a tool allowance built into their paychecks) or a division method whereby most common hand tools are bought and owned by the technician and equipment-specific tools are purchased by the company. None of methods is the perfect solution. There will always be some disagreement over those borderline tools that do not seem to fall exactly into one category or the other. If your technicians own their own tools and these tools reside on company property, be certain to inform them that they must insure their own tools.

Special tools can be a difficult budget item. Try to arrange a package deal on the necessary tools when you purchase the equipment. You have the most bargaining power at this point, and the tools should be readily available.

Although they are not really "special" tools, some organizations purchase air wrenches, torque wrenches, pressure-testing equipment, and electrical-testing equipment for their technicians. Permanently assign someone to control, organize, and inventory these tools or they may quickly disappear.

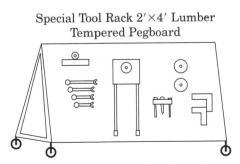

Figure 5-23   This simple pegboard rolling tool rack can be constructed with tempered pegboard, 2'×4' lumber, and four casters.

Special tools can be stored in a locked cabinet or on a peg-board. An effective technique for special tool control involves the construction of a pair of pegboard panels with 1/4" holes built on a pyramid-shaped lean-to frame with casters. The tools are outlined on the board with a permanent marker to establish their location and provide a quick visual inventory. The board can be locked in the parts room for security and easily rolled to the work area when needed.

6

Problems and Suggestions

A Simple Diagnostic Technique
Inventory Control and Logistics
Ethics
Requesting Money for Equipment
Effective Communication
Preventive Maintenance and Time Management
Respect and the Technician

A SIMPLE DIAGNOSTIC TECHNIQUE

Turf managers and technicians face challenges every day. A proven technique taken from standard industry practices can be employed to handle these problems. This technique is known as divide and conquer.

The divide and conquer method is useful for organizing and dissecting problems. The basic theory is simple. Almost every problem has several components. These components when approached in aggregate form can seem overwhelming. There are simply too many possibilities to sort out simultaneously. The secret is to segment the problem into solvable chunks. A good technician will approach engine and equipment problems this way. All the systems that might be connected to a problem are laid out for examination. If there is

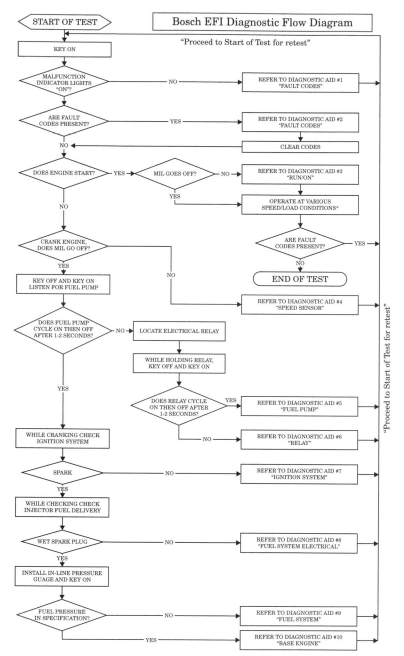

Figure 6-1 This flowchart is designed to assist the technician in troubleshooting a Kohler® CH 26 engine equipped with a Bosch electronic fuel injection system. This style of chart is a common job aid supplied by many manufacturers.

a problem with a rough running engine, the first thing to do is divide the possibilities. Is it a mechanical system problem? An electrical system problem? A fuel system problem?

Next, each of these problem areas should be further divided into discrete parts or systems. Is there fuel in the tank? If yes, is the fuel getting to the carburetor? If yes, is the fuel in the carburetor getting into the engine? If so, the fuel system is probably okay. It is time to dissect the next system because this system is working. This method essentially involves constructing a flowchart of system operation and then checking the function of each step. This same approach can be used for an irrigation problem, a turf disease problem, or even a cut quality problem. If you can become a systems or systematic thinker, problems will be solved more accurately and expeditiously. It helps to write the steps down the first time you tackle a problem since the same problems occur over and over. As the Italian mathematician Pareto discovered, the same 20 percent of your problems will occur 80 percent of the time. Be sure the solutions to those 20 percent are well documented. You will see those problems again.

INVENTORY CONTROL AND LOGISTICS

There will be no attempt here to instruct you on all the complexities of inventory control, but it is beneficial to understand some key concepts and rules. You usually do not have enough items or enough money invested in your inventory to focus on it too closely. Inventory is money on the shelf—tied up and, thus, unavailable for other uses. Since this is unavailable budget money and today's marketplace is served so efficiently by most of our vendors, it is tempting to draw the conclusion that you should not stock anything. This is an erroneous conclusion. The old cliché that time is money is also true. If your technician has to stop, order parts, and wait for them on every routine service, you will lose more money in labor than you can save by keeping empty parts shelves. There is economy in stocking fast-moving and frequently used items. If the technician is encouraged to put together kits for recurring turf equipment maintenance tasks and these parts are bought correctly, both time and money will be saved. You will need expendable nonmaintenance items that can be best bought in bulk. Weed trimmer line, two-cycle oil, gloves, personal protective gear, and even hand soap and toilet paper can be found on this list.

So, when is the best time to buy these items? Buy them when they are on sale. Every equipment distributor has a parts program, and every vendor has periodic sales, special deals, or free freight promotions. Take advantage of these reduced prices and freight savings. Most of the items you need have a long shelf life. If you have the storage space, a year's supply of toilet paper bought cheaply can represent real savings. Also stock up on oil, air, and hydraulic filters. Buy hydraulic hoses, hydraulic oil, and engine oil in bulk or take the quantity discounts whenever possible. Investigate the possibility of purchasing a hydraulic hose crimping machine, fittings, and hoses. This will enable you to make hoses on demand for your equipment; some enterprising shops have made hose construction a profit center by supplying custom hoses to other nearby turf maintenance facilities. It is possible to buy all the equipment and supplies for hose assembly in a package that reduces or eliminates the cost of the crimper and associated tools.

The most frequently overlooked way to obtain expendable parts cheaply is to bundle them with equipment leases or purchases. The distributor is anxious to make the equipment sale and if so, will be there with its delivery trucks to bring your new equipment. You can be sure parts freight will be free if it is on the truck with the equipment. This is the best time to ask for extra discounts on parts. You are usually not thinking about parts at this point, but rather are under the seductive spell of free warranty repair. The truth is that unless you buy an extensive and usually expensive service contract, you will need to do those initial or 50-hour services yourself. Why not get the parts you will need for those services now? You may need special greases or oils for these services. Hydraulic filters will need to be changed, and the cost of these filters alone for your new fleet will surprise you. Be sure to take advantage of the benefits available at sale or lease time.

No discussion of inventory is complete without some comments about inventory control. If you do not have your parts inventory tracked and access controlled, your shrinkage will be severe. *Shrinkage* is a term that covers the problems of inventory that is lost, incorrectly located, or stolen.

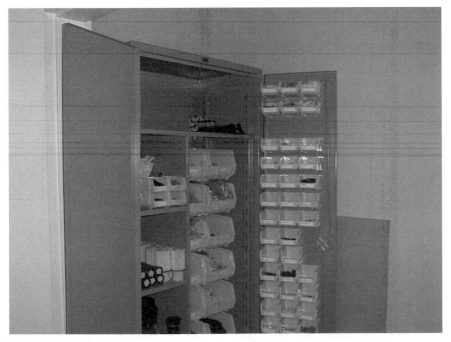

Figure 6-2   This is an example of an enclosed parts cabinet.

Figure 6-3   All parts in this parts room at Old Collier Golf Club in Naples, Florida, are in covered storage.

Figure 6-4   A low-budget parts room—being neat and organized has zero cost.

Many people hear the term *inventory control* and envision armed guards, extensive and time-consuming tracking procedures, and vaultlike enclosures for all the parts and supplies. Most of this is not necessary. Always follow the rule that you need no more security than the worth and desirability of the item being protected. A case of 5-W-30 motor oil next to the side door of the building may disappear quickly, but a case of mole bait may not. High-value, commonly used, and easily sold items are tempting.

Many facilities have found that two levels of security are effective. Those high-value parts and lubricants need to be under lock and key with careful accounting. The items needed every day out on the course do not require as much security or accounting. Soap, hand towels, toilet paper, weed trimmer line, ear plugs, and similar items may be best kept in open inventory. Simply put them on a shelf easily accessible to the crew members who need them. When you implement this type of system, keep a careful inventory for a month or two to satisfy yourself about the shrinkage rate. Suppose you lose about $100 a month with this open inventory method. You may find you cannot afford to keep the items under lock and key if you calculate the shrinkage against the expense of control. Setting up an employee in a one-hour-a-day "store" to dispense expendables is not smart. It makes no sense to pay someone $8 an hour for 20 hours a month to prevent a $100 per month loss. Remember, too, that the employee costs much more than $8 an hour. Just let the $100 per month loss go.

The logistics of obtaining everything from fertilizer to safety glasses can be challenging. Realize that if you are in or near a large metropolitan area, your problem will usually be too many suppliers, not too few. This can be as big a problem as the logistics in an isolated area. You are inundated with offers. They come from salespeople, catalogs, flyers, and emails. You lose tremendous amounts of time just handling communications. If you have too many suppliers, you will overcomplicate your life. Simply cataloging and filing invoices can be a challenge. Putting all these people into your accounts payable system is painful, and the possibility of errors, missed payments, and receiving the wrong merchandise multiplies. Make your suppliers few and well chosen. Always require them to make appointments that fit your schedule. Actively discourage walk-ins. Pick suppliers for their problem-solving abilities, not just their prices.

A collaborative salesperson who works with many other similar customers in your area has a wealth of information to offer. If you keep a list of questions alongside your want list for the supplier and your salesperson fills both your want lists, you have a successful relationship. These relationships are especially important to technicians and equipment managers. If you create a rapport with your distributor's service, parts and salespeople, you will reap many benefits. You are much more likely to get exceptional service, heads-up advice on technical problems, and money-saving offers from your distributor if you are liked because you are pleasant, flexible, and friendly. It may seem that the squeaky wheel gets the grease in many circumstances, but experience indicates that pleasant, no-hassle customers get extra services and primary consideration. As many smart managers have discovered, being pleasant is a zero-cost way to reap many benefits.

The more accessible your supply chain is, the fewer parts you will need to stock. This seems obvious but needs to be explained. There are several caveats to remember as you research your primary nearby suppliers. The number one item to discern is this: does your supplier have a broad inventory of the things you need? If your supplier stocks only parts for the latest models of machines and your fleet is six years old, then you are going to have to increase your parts stock or live with lengthy downtimes. Check your distributors' fill rate on orders. Realize, too, that the elapsed time from ordering to arrival at your facility is also critical, as is the completeness of the order. An order fill rate of 100% is meaningless if you need two bearings with the same part number to repair a machine and receive only one. All these considerations influence what you must keep on your shelves.

Stock preventive maintenance (PM) items by creating kits. The kit concept can be explained as a bundle of parts that are required for a particular service on a machine. A 100-hour service on a light fairway unit may require a hydraulic oil filter, an engine oil filter, an air filter, spark plugs, and a fuel filter. Create a kit that contains all these parts by boxing or bagging them and creating a unique part number, such as JacLF1800hr100. When this service comes up on your PM chart for this unit, neither ordering nor looking up and pulling all the parts is necessary. One quick check in one location and you are ready for the service. Remember that these are the parts you should buy on distributor promotion or at a special price when you purchase new equipment.

Tips on Parts Stock and Vendors

- Supportive vendors and suppliers are a tremendous asset. Cultivate these relationships. They will do much more than supply parts.
- Stock the parts necessary for preventive maintenance (PM) services. Create kits to save time.
- The depth of your inventory is contingent upon the length of your supply lines. If you get dependable, overnight service from your distributors, your inventories should be minimal.
- Create only as much security as your potential losses deem necessary. Too much security is as costly as too little.

ETHICS

Ethics in the turf facility management business can be a controversial and sometimes elusive subject. There are countless ways that ethics can be subverted. The equipment manager who changes the oil in his personal vehicle on company time is stealing, yet this is not an uncommon practice. The superintendent who has his equipment manager rotate the tires on his personal vehicle is also stealing, and again this is a far too common practice. The facility manager who "borrows" a rake, a bag of seed, and a bag of fertilizer to use at his home is stealing. This particular event was observed moments after the thief was heard disparaging some of his crew members for taking toilet paper from the supply cabinet and putting it in the trunk of their cars. This manager was genuinely disgusted at the "thieves" he had working for him and never suspected his actions were contributing to the problem.

Ethics is partly about modeling behavior. The facility or equipment manager cannot expect or demand that his or her employees stop stealing until he or she stops. The behavior modeled is always more powerful than the words coming out of the manager's mouth. A scrupulously honest manager can make demands that are credible. Always try to create an atmosphere and environment that promotes honesty. This atmosphere is created by your actions as a manager. The environment is created by reducing temptation. Good inventory control and secure storage are essential. Reducing or eliminating traffic through the shop and storage areas is a great way to control theft. Keep the fuel pumps locked, and supervise all fueling activity. The ideal situation is to keep personal vehicles out of the shop, equipment storage, and fueling areas. Creating a "No Personal Vehicles" zone around the shop and storage areas greatly reduces many temptations.

REQUESTING MONEY FOR EQUIPMENT

Worn-out equipment is very debilitating and hampers cut quality, production, and crew morale. Equipment, even with the best care, wears out. There is a point when major repair or rebuilding cannot be justified. Replacing engines and hydraulic pumps is usually not cost-effective on high-hour machines. The better records you keep, the more probable it is that you can present an effective case for replacement. Sometimes, your customers are your best allies in your fleet replacement efforts. When customers are complaining about conditions and cut quality cannot be maintained because of worn equipment, let this fact be known. Cultivate good rapport with your most influential customers. Let them know why the cut quality or inability to keep the turf maintained is occurring. Sometimes, a request for additional repair funds will precipitate an inquiry from management about equipment condition. More frequently, you will have to build a case for your request. If you have good maintenance records, you can vividly illustrate why a piece or fleet of equipment is no longer viable. If the monthly repair on a unit exceeds a monthly note on a replacement unit, the math can be made obvious to everyone. If you need or can afford to replace only select units, these records will again be your ally. You will know the lemon or most expensive-to-maintain units. These are the units to target for replacement. Gather your facts carefully. Anticipate questions and build a list of responses.

Be prepared to "buy down" your request, if necessary, with fewer or less expensive units. If you have a low-budget operation, preprice good used replacement equipment. There are operations with generous budgets that roll over low-hour equipment every few years. A well-maintained, previously owned unit will serve you well and save money. One operation in southern Florida that is known for literally overmaintaining its equipment holds an auction every few years and draws a large crowd of bidders. The operation's reputation for selling off equipment that is as good as new serves both it and its customers well. It commands top dollar, and its customers get superior used equipment.

EFFECTIVE COMMUNICATION

Many managers complain that their crew members just do not listen, or do not understand or do not get the message. Far too often, the real problem is that the managers are not communicating effectively. Using lingo unfamiliar to their employees or talking over or even down to their employees is a common problem. Miscommunication can be costly and even life threatening. The cost of doing a poorly communicated task over is usually costly, but in some instances, it can be fatal to your career. The following is an actual example of a communication failure. A spray technician who worked at a soccer complex was a smokeless tobacco user. His assistant was an H2B visa worker. The spray technician had been instructed by his boss to spray a selective herbicide on the 40-acre main field. Through a wad of tobacco, he tried to tell the assistant what to spray. The assistant misunderstood his instructions, sprayed the wrong chemical, and killed all 40 acres of grass. Who was at fault? The spray technician was fired, and the assistant reassigned. The tobacco was, in effect, extra "noise" in the communication process. Noise is one of the many enemies of good communication. Noise can be in your physical surroundings (e.g., a fan or equipment running nearby) or even your receptors' mental state (e.g., the receptor of your communication is thinking about the fight he had with his wife before work).

Tips for Effective Communication

- Eliminate noise whenever possible. Seek a quiet place for communicating. Turn off fans, equipment, and other physical noisemakers.
- Do not finish other people's sentences, even mentally. Try to quench your personal opinion of the speakers, and actively listen to the message instead.
- Learn to become a body language expert. If the message is mixed, trust the body language first, but be careful. Folded arms can mean an unreceptive listener, but they also can mean a chilled listener!
- Employ appropriate communication methods. You would not send an email to alert everyone that the building is on fire. Make the message and the medium match. Match the medium to the listener, too. A poor reader needs verbal communication.

There are a number of other disruptors of the communication process. Language barriers have been mentioned. Cultural differences can hamper communication. We also sometimes deliver a mixed message. Our body language says one thing and our voice another. Often, there can be long-standing distrust between the two communicators that hampers effective communication.

When considering the communication process, we sometimes overlook the importance of listening. Learning to actively listen is difficult, and while hearing is easy, it is more difficult to listen. We all have a tendency to mentally finish other people's sentences or mentally cut them off as we plan our response. In this process, we stop listening. Learning to listen is a skill that requires continuous practice.

PREVENTIVE MAINTENANCE
AND TIME MANAGEMENT

If you have questions about the basics of PM, please see Chapter 3. This section focuses on the impact of PM on time management.

Many managers have never tried to initiate a PM program. They are completely occupied with day-to-day concerns and simply keeping most of the equipment limping along. This is a time management disaster. Contrary to what you might think, PM will reduce panic repairs and late-night repair sessions. It will not appreciably reduce your overall workload. This is a myth that leads to disappointment and disaster. Equipment managers will state that they tried PM, and it was just too much work or did not work. Undeniably, it is work. It demands organization and a lot of extra work during the implementation stage. If you begin implementing PM in a "run until it breaks" shop, your workload may momentarily double. You will have to arrange extra hours and, perhaps, additional employees in the shop on a temporary basis. You will be in essence running two systems simultaneously. This guarantees extra work will be necessary.

However, what you will gain as you slowly bring more and more equipment under the PM program is a predictable workload and, consequently, a much more manageable workload. You will have a look at the work ahead for several months with a good PM plan. This organization will allow you to manage your time more effectively and plan for upcoming events. You will no longer have panicky moments or run up huge overtime bills just before aerating, overseeding, or getting ready for a tournament. You will know your aerators, top dressers, spreaders, and allied equipment are ready because you got them ready and keep them ready under your new PM plan.

RESPECT AND THE TECHNICIAN

Why would an equipment technician deserve respect from his or her boss? Why should the crew respect the technician? Quite often, there is a lack of respect for the technician and the technician's position in the organization. There are several almost invisible but essential functions that a technician can perform effectively only in an atmosphere of trust and respect.

We all can appreciate a technician who keeps the equipment in good working order. Most managers appreciate this situation, especially if they have ever worked with a less-than-competent technician. The technician who is respected by management will often also be respected by the crew. Respect from the crew is more essential than you might realize. The crew and the technician have a unique relationship. Good technicians have a fierce affection for their equipment. They want it to work perfectly and to be operated correctly. Good crew members appreciate this fact. Their job satisfaction depends on equipment condition. If the technician and the operator have a good rapport and mutual respect, then troubles are few. An operator who damages a piece of equipment or detects an unusual noise knows that that the technician needs to be alerted immediately. If there is mutual respect, there will be no recrimination. Defects and damage are not hidden from the technician but, rather, promptly pointed out. The operator is the expert eyes and ears of the technician for the piece of equipment he or she operates. The technician is working blind without this line of communication. This communication must also occur between the technician and his or her supervisor. The technician has volumes of information about crew morale, the abilities of individual operators, the opinions of the crew members about each other, and the crew's opinion of the supervisor. This is valuable information that will not be communicated in an atmosphere of disrespect and distrust.

If you manage a technician, you must treat him or her as a valuable team member. Technicians cannot be called "grease monkeys" or other disparaging terms behind their backs. This disrespect will not be hidden from the technicians. The difference in education levels and social backgrounds can be difficult to ignore, but good supervisors will put these concerns aside. There is too much for the supervisor to gain from a congenial, communicative relationship with his or her technician. A good relationship can mean that the technician serves as a neutral third-party liaison between the supervisor and the crew. A good technician will know when the crew has problems and issues with management. Often, the technician is the first to know. Work with your technician. Put yourself in the technician's shoes. You will definitely benefit from a well-cultivated partnership. Do not forget this old cliché: Good technicians make mediocre supervisors look good. Bad technicians can cost good supervisors their jobs.

Training Equipment Operators

Where to Find Good Operators
Training Equipment Operators

WHERE TO FIND GOOD OPERATORS

Equipment operators are essential members of the crew, and their equipment gets the most use. Equipment operator positions are usually considered entry-level positions. Entry level does not mean that these employees are easy to find or that anyone hired can be instantly proficient at mowing.

Finding dependable employees is the biggest challenge in any business. In the past, high school and college students would frequently seek part-time outdoor jobs. Most of these job candidates had been inside a classroom for a large part of their lives and sought the change in environment. These potential workers were drawn to various outdoor part-time jobs, including golf course maintenance and sports turf work. This phenomenon is not as common among young people today, but the situation is not hopeless. There still are several potential sources for part-time and full-time crew members, although the hourly wage, full-time, dependable employee can be difficult to find. Some ideas for sources of employees are listed below.

High school and college students: Some students still want to work at golf courses where golfing privileges can be part of the benefits. Advertising at a nearby college remains a good idea. With ever-increasing gasoline prices, a job at a golf course or sports turf facility close to the student's college or school may interest him or her. Building a rapport with counselors at local community colleges or high schools may yield part-time workers, and such work may help a student discover a turf-related career. A high percentage of community college students work part-time while attending school. Luring a high school senior into part-time work can result in a good employee for several years or even produce a permanent employee. A number of sports turf and golf facility managers have discovered that offering a scholarship in the turf field to one of these students can be most productive. The student is awarded the scholarship with the contractual agreement that he or she will work for the facility for a year or two after graduation.

Retirees: In states like Florida, South Carolina, Arizona, Texas, and California, where people flock to retire, this population can provide outstanding part-time help. Even retirees who had senior management careers are potential candidates for equipment operator positions. Those who are in good health and who want to stay active, who like being outside, and who play golf often like the idea of mowing early in the morning at a golf course a few hours a week in exchange for golfing privileges. Some golf courses have been very successful using this locally available talent. Retirees who had high-pressure careers often like the peaceful, low-pressure environment of mowing. Using several retirees can fill your crew positions with very dependable people. While this employee resource may not be available to every golf course or sports facility, it is worth exploring.

Workforce boards: Most states have workforce boards or something similar as part of national welfare reform. These boards help people find jobs, retrain unemployed people, and often provide funding for training. A golf course or sports facility can register with the local workforce board, indicating how many jobs they have open and listing the starting wages. People who walk in for help could be made aware of these jobs. Building a good rapport with workforce board personnel can be helpful.

Foreign workers: Doing an Internet search for "legally hiring foreign workers" or a similar term will yield various resources for hiring international employees. Much of America's turf industry workforce is Hispanic, but workers from many other countries can also be hired legally through the government H2 visa program. Some of these workers can be hired on only a temporary basis, and they must return to their country of origin at specified intervals. Many Internet sites list contractors who coordinate the arrangements, but the hiring company's attorney should also be involved in this process. International workers can be a great resource for hard-working, dependable employees, but there are challenges. Language differences are obvious. Sometimes even employees who speak Spanish but came from different countries may have cultural differences that prevent them from getting along. It can be a real challenge for management to deal with such differences even if the supervisor speaks some Spanish or has a good interpreter. Many golf courses are hiring foreign workers, but due diligence should be taken to ensure they are here legally, and management must consider the variety of challenges in advance. There may be local seminars available to provide helpful information on hiring and managing international workers. It is desirable to be well informed on the proper management of foreign workers before deciding to use this potentially valuable source of employees.

TRAINING EQUIPMENT OPERATORS

Training equipment operators can be approached in several ways. Traditionally, operators were taught by other operators: one of the better operators would show the trainee how to operate the equipment. This could be disastrous if the trainee did not "get it" when the master operator demonstrated his or her craft. Several things usually go wrong in this scenario. The assumption that is frequently made is based on the belief that showing and telling is teaching. Nothing could be further from reality. The second most frequent mistake is assigning the best operator as a teacher. Far too many expert operators do not know how they do what they do. The concept is known as the unconscious performer phenomenon. They just "do" what they do. If they do not know how they do it, they cannot teach anything to the trainee. A more productive approach to operator training should begin by selecting a permanent trainer. Select the operator or other employee to do all your training on whether he or she:

Is friendly and well liked by most crew members
Is patient and a good communicator
Understands that trainees learn at different rates and in different ways
Is willing to teach
Understands the importance of the trainer's job

Is willing to accept meaningful compensation for this additional work

You must also consider:

If you are willing to train the trainer.
If, better yet, you are willing to have a professional educate your trainer about the basics of teaching

If you can answer yes to these criteria, you can create a trainer who will teach your employees quickly, safely, and competently.

Once a trainer has been selected, arrange for him or her to be taught the basics of teaching. Look at your local community resources, workforce training centers, colleges, universities, and large industries in your area, or anyone who trains large numbers of workers. They have found ways to train their trainers. Learn from these training professionals and training centers. You may already have a training resource person in your organization. The largest training outfit in the world is the U.S. military, which has taught soldiers how to teach for more than 100 years. A retired Army or Air Force sergeant or Navy chief can often teach the basics of training to your new trainer.

This may sound like a lot of effort to expend to create a competent trainer, but it will be effort well spent. The better you train your employees, the less money they will cost you in damaged turf and equipment. The employee will be happier and less likely to go elsewhere, thereby reducing turnover. Remember that you have two choices: you may train an employee and he or she might leave, or you might not train him or her and he or she might stay. A well-trained employee is a good investment.

Once you have a good trainer, what should your trainer teach? Your first, best resource is the manufacturer's operator manuals, videos, and other such information. You need to assemble this information in a series of drawers or bins, with each labeled with the manufacturer's model number. Create a training checkoff list or wall board that indicates the training level of each operator. The operator, you, and the rest of the crew will know who is trained on which machines. Your policies and operating procedure manual should reflect the necessary competencies for each crew position, including which machines must be mastered for each job and/or job level. Many successful training plans tie machine and other job mastery to pay and promotion. The more operators you have who can operate all the equipment competently, the better you will sleep at night. Cross-training is beneficial to everyone.

Policies and Operating Procedures Manual

A well-designed policies and operating procedures (POP) manual is a superb management tool for the turf equipment facility. It should contain job descriptions for all positions, a complete list of all company rules and policies, and most important, an exhaustive job description for each position. When creating this document, do not feel limited to a specific number of pages for each job description. Be sure to include digital photos of correctly completed tasks where possible. Realize that the more detailed and comprehensive you make this book, the fewer problems you will have with "it is not my job" arguments and issues with how well, how long, and how much. Quantifying and detailing each position's requirements reduces stress for everyone. This is also the place to detail prerequisites for promotion and pay increases. For best results, tie them to training competencies.

8

Filling the Turf Equipment Manager Position

Lack of Career Awareness
High Schools as a Resource
Postsecondary Programs
Succession Planning

LACK OF CAREER AWARENESS

It is clear that the turf equipment management position at a golf course or sports facility has become an increasingly important position. It is a career path with plentiful jobs, excellent starting salaries, and lucrative advancement potential, but little recognition outside the golf industry. It is incumbent on the turf industry to get information to people about the various careers in the turf industry, including the position in short supply, the turf equipment manager. Building a rapport with a local high school, radio advertisements, and pamphlets are a few methods that could be used to inform the general public about this career. Lack of career awareness is a major factor in the low supply of turf equipment managers.

HIGH SCHOOLS AS A RESOURCE

Where is the turf industry going to get the needed supply of properly trained equipment technicians who are mechanical; who know hydraulics and electrical systems; who know reel sharpening, setting, and repair; and who are shop managers? Mechanics from the automotive and other industries could convert to the turf industry, but there is a shortage of all mechanics, so it does not seem that this group will fill the void of turf equipment managers. It seems more logical to recruit young people who are searching for their first mechanical career.

High schools can be a source of turf equipment technicians if the students know about the career. Sports turf managers and golf course superintendents should be proactive by implementing a high school career awareness program. A good start would be to call a local high school, and talk to the counselor and agriculture and/or automotive mechanic teacher. Invite them to the maintenance facility, and talk to them about all that the turf industry has to offer. Seeing a neat, organized turf maintenance facility and learning all that is involved in turf management would be a tremendous educational experience for them. Educate them, perhaps provide them with lunch, and, generally, make them feel good about the turf industry. Next, invite these educators to plan a field trip to the golf or sports turf facility so the students can be enlightened. Proactively reaching out to the high school is the beginning of developing a long-lasting rapport with educators.

It may take some reminder calls to the high school to get the student field trip arranged, but be ready to accommodate the students when a trip does occur. Have the maintenance facility clean, have equipment available to demonstrate, and have the turf management team members talk about what they do and how they discovered their career. Just a few hours in the shop seeing all the specialized turf equipment and what is involved in keeping it maintained may be enough to spark career interest among the student group.

The next step in developing career interest is getting some experience at the job. With the student group there, the turf manager could invite high school seniors to apply for part-time jobs. Thus, the student gets an opportunity to experience the turf industry, and the superintendent or sports turf manager might discover a student who has a good attitude and work ethic that he or she wants to keep. Follow this with a scholarship from the club or from the local or state turf association if the student attends a turf equipment management program and returns to the club or club's chapter area upon graduation.

By developing a high school career awareness program, the turf manager could help students discover careers, hire some local part-time employees, and potentially find a future full-time staff member. There could also be the added benefit of receiving some positive public relations by helping local youth. This can be a win-win situation that makes the high school a valuable local resource for employees, especially for difficult-to-fill positions like the turf equipment manager.

POSTSECONDARY PROGRAMS

Postsecondary programs include vocational-technical centers, community colleges, and colleges and universities. All of them may be sources of part-time employees, especially if they are near the club or sports turf facility. The vocational-technical centers and the community colleges are also the most likely places to find full-time equipment managers, particularly if the school has any mechanical-related programs.

If there is a technical center or community college nearby, it is worth the superintendent or sports turf manager taking the time to learn the school's program offerings and to become familiar with the instructors involved. Reaching out to the instructors may develop into a relationship that yields part-time employees and, perhaps, a new turf equipment program at the school.

Colleges and technical centers have counselors and job placement personnel who are also good resources. As discussed earlier, in reference to high school counselors, getting to know these people and wooing them makes sense. The more these educators know and understand about the entire turf industry, the better. With such knowledge, they can recommend part-time jobs at the nearby golf course or sports field to students. The counselors also work with community college students who are seeking an associate in arts degree with the intent of pursuing a bachelor's degree at a university; however, the students may not know which major to declare. Exposure to a golf course or sports turf facility may lead them to major in horticulture, plant science, or even mechanical engineering.

Serving on an industry advisory board for a local horticulture or mechanics program provides another opportunity for a superintendent to become familiar with the program content and the students. Typically, this might require one or two meetings a year; as such, it usually does not require a weekly or monthly time commitment. Through this involvement, the superintendent might hire students for program internships or just as part-time help.

Community colleges and vocational-technical centers respond to the needs of local industry, so a superintendent who gets involved with a local college could suggest that a turf equipment program be initiated. Doing so would require superintendents in the area to collectively support the new program not only by agreeing to hire graduates, but also by providing scholarships and committing to spending some time to make high school students aware of the turf equipment center. There have been superintendent-initiated turf equipment programs that were closed within a few years of starting because of low enrollment. So, simply agreeing to hire graduates and offering scholarships, although very important, are not enough to keep a turf equipment program viable. Enrollment is a critical factor, and it is not easily achieved, a result of a lack of awareness of the turf equipment career.

SUCCESSION PLANNING

Commonly, a golf course superintendent or sports turf manager is not concerned about the turf equipment manager position until he or she has the position open. Then it quickly becomes a near emergency to fill the opening since the equipment manager is so important to the entire operation. Sometimes, the equipment manager leaves on short notice, which is difficult, but some planning can ease the situation.

Succession planning refers to thinking in advance, "What will I do if my equipment manager leaves suddenly?" It is common not to consider this situation if one has a good equipment manager since finding a replacement can be trying. At Lake City Community College (Florida), we have received calls from superintendents almost in panic mode because the equipment manager left suddenly and the superintendent was having difficulty finding a competent replacement. Planning may not make it easy to find a replacement, but preparing in advance for this loss can make filling the position easier.

The ideal situation is to have an assistant technician who can work with the head turf equipment manager, so someone is being trained to take over when needed. The assistant could be someone on the crew who is mechanical or someone who is hired specifically to fill the assistant technician position. Students graduating from equipment programs should first work as assistants under a good turf equipment manager, so they can hone their skills and gain experience under real conditions. Not all clubs or sports turf facilities have an assistant technician position, but it is very desirable; it is good for the facility and for the beginning technician. Having a trained assistant who could take over is a real advantage. Searching for another assistant is less stressful than scrambling to replace the head equipment manager with no replacement in mind.

The assistant technician should be trained by the equipment manager, sent to seminars and factory and distributor training sessions, and, perhaps, sent to a local community college for courses in mechanics, computers, and management. The more that is invested in training the assistant, the easier his or her transition to equipment manager will be when the need arises. This situation takes a lot of pressure off the superintendent or sports turf manager since there is comfort in knowing that a replacement for the equipment manager is being trained.

There are some who would not want to train an assistant technician for fear that once the assistant has good training, he or she will leave to become a head equipment manager somewhere else. That certainly is a risk as it is with all positions, but it should not preclude one from building a well-trained team. Treating people with respect and providing a work environment conducive to employee loyalty is the best insurance against turnover. Having a well-trained assistant technician is a good insurance policy.

If turf equipment managers are in such short supply, how can a turf manager find an assistant technician? High schools, vocational schools, and community colleges can be good sources. As stated earlier, building a rapport with the instructors of any mechanical-oriented program at local schools can yield a supply of part-time help for the shop and for the turf maintenance crew. It is desirable for an equipment technician in training to operate equipment, mow turf, aerify, etc., so he or she can better understand the equipment and can communicate with crew members who come in with an equipment complaint.

At Lake City Community College, most of the Turf Equipment Technology graduates need to work for a year or more under a good turf equipment manager. Graduates are often offered head equipment manager jobs because the demand is high and the supply is very low, but this often is not the best for the student or for the facility, and we discuss this with our students and employers. We do have some experienced graduates who could tackle a turf equipment manager position, but this is not the norm. A golf course superintendent or sports turf manager would really help him- or herself by developing an assistant technician position and proactively training that person. It is a wise investment in the future to have an assistant technician in training ready to replace the head turf equipment manager.

Another way that golf course superintendents and sports turf managers can find assistant technicians is to initiate a "feed the chapter" program at the local turfgrass superintendents or sports turf chapter. *Feed the chapter* is a term used to define a career awareness program that could be developed by the various turf-related industry chapters in any area. Basically, "feed the chapter" involves the following steps with modifications to meet local needs:

At a chapter meeting, ask for volunteers who have neat, clean, organized maintenance facilities to contact a local high school.

Contact the principal and explain who you are and that you want to invite the principal and the appropriate counselors and teachers to your turf maintenance facility to discuss careers and, especially, the national shortage of turf equipment managers.

Set a date for the visit, and when they arrive have the facility in top condition. Discuss all that goes on in the facility, the expense of the equipment, the need for competent turf equipment managers, etc. This provides a great opportunity for the golf course superintendent to outline all that is involved in implementing modern best management practices for golf courses and how concerned the golf industry is about the environment, wildlife management, and water use management. This visit can be a great learning opportunity for these educators. Then provide lunch, and have them return to the high school with a new understanding of the golf and sports turf industry and, ideally, excited about having their students involved in the industries.

After the educators' visit, explain that you would like to do something similar for a select group of students. Many high school principals, counselors, and teachers would be thrilled to learn of a way to introduce their students to new careers. School administrators get busy just as superintendents and sports turf managers do, so you may need to make follow-up calls to ensure the student visit happens.

When the students visit, again make sure the facility is in excellent condition. Have the equipment manager, the assistant superintendent, and any college interns make presentations to the students about how they got into the turf industry and emphasize the shortage and importance of the turf equipment manager. This can be fun for the golf course or sports turf staff, and it will be a real eye-opener for the students. Provide lunch for the students, if at all possible, and send them back to the high school with an understanding of the turf industry and the various career opportunities.

Initiate a part-time worker program for the students from the high school you contacted. Now that the teachers, administrators, and students know about your facility, getting students interested in part-time work should be easier.

Give some students part-time work, and hire some seniors in the summer after they graduate. Pick out the students who have a strong work ethic and good attitudes, and approach them with an offer of a scholarship from your turf chapter to attend a turf equipment program if the student agrees to return to work in your chapter area.

If just a few members of a golf course superintendents chapters or sports turf chapters in every state would get involved in this program, think of all the students who would be made aware of this mechanical career and other turf careers. Many golf course superintendents chapters raise money to send students to turf programs, so why not earmark some of that for turf equipment programs?

The other advantage of the "feed the chapter" program is that the chapter would be helping local students return for local jobs. Most communities consider this a positive situation. The local golf course superintendents or sports turf chapters would gain some positive local press by getting the scholarship program with recipient names mentioned in the local paper. This could be welcome news about the golf course and sports turf industries, which are often misunderstood by the public.

9

Safety Regulations and Regulatory Agencies

Clean and Organized Is Safer and More Efficient
Assistance from Regulatory Agencies and Your Insurance
  Carrier
Handling Fuels and Oils
Chemical Handling
Recycling and Detoxification
Hazardous Materials Record Keeping
Accident Prevention
Accident Reports and Reporting
First Aid and First Responders

CLEAN AND ORGANIZED IS SAFER AND MORE EFFICIENT

What takes more time at your facility? Finding the tools to fix the problem or fixing the problem?

Most of us waste a tremendous amount of time just trying to find the tools and materials to take care of a task. The irrigation repair tools are piled in the back of the utility bed amidst the dirt and scraps of pipe, empty cans, and assorted trash. You get to the leak, which is usually a half mile from the maintenance facility, and discover you do not have the fitting you need. You must have used your last one on a previous repair, but in that jumbled mess, how would you know? After several abortive efforts and two or three trips to the maintenance facility for materials, you have the job done. Unfortunately, it took one and a half hours to complete a 15-minute job. This same scenario is repeated in the shop and on the course day after day. You are working 60 or 70 hours a week and not getting everything done. What is the problem?

It is obvious you need to get organized, but who has time for that? The real question is, who does not have time to get organized and neat? In any facility where tasks are repeated almost daily, there is no excuse not to be and stay organized. It is merely a matter of economics and time management. You cannot afford to hunt for tools and materials. If more than one person uses the same tools and materials, it becomes even more critical to have a place for everything and a habit of returning things to their rightful places.

There is a multitude of benefits to clean, neat, and organized surroundings. You are many times more efficient when you are organized. Your workers or coworkers are much more efficient when tools or materials are being shared. It is a boon to productivity when everyone returns a tool to the same location after each use. Safety is a big concern. Tools scattered on the floor are a slip and trip hazard. Technicians with injured backs from falls and other accidents are far too prevalent in our industry.

There are plenty of solutions to common storage problems. The utility vehicle used for irrigation repairs can be organized in a very sophisticated way with wooden dividers and covered compartments to store tools and materials, but this is not the only way to organize. Six or so five-gallon buckets can create a quick organization system with little expense and effort. It may not look elegant, but tools and supplies organized in six compartments beat everything in one tangled pile.

Another effective solution to storage of shared, frequently used tools is a small roll-around tool cart. This style of cart is handy for organizing all your reel height adjustment tools in one portable unit. Other uses for this type of cart include a unified storage location for reel grinder stones and accessories or even a place to store hand lubrication equipment. This mobile cart can be rolled into the parts room or small equipment room for overnight security.

Staying organized is a continual process. If you think you can get organized once and be done, you have misunderstood the concept. Once you are convinced that neat and organized is an indispensable philosophy with good payback, you will succeed in being organized.

Figure 9-1   Matco locking tool cart.

ASSISTANCE FROM REGULATORY AGENCIES AND YOUR INSURANCE CARRIER

The Occupational Safety and Health Administration (OSHA), determines the law of the workplace in the United States, with some interesting variations. There are currently 22 states that have their own plans. These plans must be "at least as effective as" the federal standard. You can find a list of these states at www.osha.gov/dcsp/osp/faq.html#oshaprogram.

In some states, cities and municipalities fall outside the OSHA rules. If you work for a city or county government, investigate the variations of these rules as they apply to you.

In order to become familiar with these rules, you can visit www.osha.gov and search for topics. Also be sure to take advantage of the training offered at the site under "eTools" and "eTools" in Spanish. You can generate reports and use the electronic "Expert Advisor" tools to analyze your local conditions. You can get a feeling for the regulatory structure of OSHA at this site. Additionally, it is suggested that you obtain a copy of *Keller's Official OSHA Safety Handbook* and decide whether you want to utilize this book to train your employees on the basic OSHA rules and procedures. Training is a required and continuous component for OSHA compliance. This book is inexpensive and easy to read and understand. It is a handy training aid for introducing employees to OSHA regulations, safety topics, and fire prevention.

Your insurance carrier may be a good resource for maintaining a safe, efficient shop. Many insurance companies will provide an inspection of your facility and make recommendations. Your insurance carrier also has the right to make these inspections unannounced, but, undoubtedly, it would be happier to be invited to assess your risks. If you are averse to inviting your carrier to inspect your premises, you can hire an independent inspector to do a safety audit for a fee. Some of these inspectors specialize in OSHA compliance. Typically, these inspectors are underwriters who contract with insurance companies to make inspections for them. These inspectors are usually quite adept at assessing slip, trip, and fall issues; lifting; ladder safety; fire prevention; and other common industrial risks.

Another source of risk assessment is your local or state fire marshal's office. Its primary focus will be on fire safety, fire extinguisher locations and types, and other fire protection–related issues. It is also a good idea to have your local fire department familiar with your operation and which fuels and chemicals are on the property.

An often overlooked resource for safety audits of your facility are your vendors. Your fertilizer, herbicide, and pesticide distributors are experts on what they sell. These vendors are familiar with the regulations in your area and can make suggestions for not only correct application, but also safe handling and storage.

In some areas of the country, the local technicians associations have an annual meeting focusing on safety, with an emphasis on the latest U.S. Environmental Protection Agency (EPA) and OSHA rules. Often, trainers and inspectors from the local offices of these agencies are in attendance. This can be a great, low-stress way to get updated on the rules affecting your locale.

HANDLING FUELS AND OILS

Gasoline, diesel, and hydraulic and lubricating oils have specific storage and handling rules. The rules vary from state to state, so only the basics will be covered here. Your state department of environmental protection (there are more than 20 states that have their own agencies) and your state fire marshal will likely have a website you can access for specific rules. Some states follow EPA rules and others use modified EPA standards, but usually, the EPA standards are the minimum rules.

In areas where the water table is near the surface, gasoline and diesel storage tanks are located aboveground. Because the whole state has a very shallow aquifer, Florida and local county rules in Florida demand aboveground tanks in most new installations. The most effective way to handle these rulings is by using double-wall tanks. Single-wall tanks require impermeable containment walls that are 110% of tank capacity, and the tanks need to be covered to minimize evaporation losses. The use of a monolithic, double-wall tank solves a number of the problems associated with single-wall tanks. Most double-wall tanks have a simple, built-in leak detection system. These vault-type tanks are insulated against evaporative loss and use a mass of concrete on the exterior (some with an attractive pebble finish), which

Figure 9-2   A ConVault® tank at Golf Club of the Everglades in Naples, Florida.

provides bullet and crash resistance. You should also investigate the local rules for your specific tank on size versus regulation. In some states, tanks below a capacity of 1,100 gallons have less-demanding design and installation rules.

Used oil must always be correctly labeled as "used oil." "Waste oil" and "recycled oil" are not acceptable labels. Many facilities have found that a small double-wall vault works well for temporary storage. The EPA is not as strict about storage containers for used oil as they are for fuels, but proper labeling is mandatory. There are a number of used oil removal services throughout the country. Be certain they have an EPA license number, and get a receipt for your used oil. If you generate fewer than 25 gallons of used oil a month, you may be exempt from classification as a "generator" or you may be classified as a "small generator." Be sure to check state and local regulations on used oil handling.

Used oil filters have handling guidelines that vary from state to state. Call the Filters Manufacturing Council at 1-800-99FILTER to obtain the rules for your state.

CHEMICAL HANDLING

The heart of the OSHA is the HAZCOM or "right to know" portion of the law. These standards deal with your employees' right to know which chemicals they are exposed to in the workplace, training plans for handling hazardous chemicals, proper labeling, and a plan to relay this information and training to your employees. You are required to keep a log, accessible to all employees, of all the chemicals in your workplace. This log must list each product/chemical and follow a prescribed arrangement. The hazards, cautions, chemical composition, and other data must be listed for each. There are some minor exemptions for everyday products like hand soap, lotion, copier toner, and similar products. When in doubt about a product, find the material safety data sheet (MSDS) on the Internet, print a copy, and post it in the log. Keep your log up to date. OSHA or your state agency will check it during an inspection. You are also required to have an up-to-date training plan, records of training sessions, and complete documentation of your labeling efforts.

Material Safety Data Sheet

SECTION I - Material Identity
SECTION II - Manufacturer's Information
SECTION III - Physical/Chemical Characteristics
SECTION IV - Fire and Explosion Hazard Data
SECTION V - Reactivity Data
SECTION VI - Health Hazard Data
SECTION VII - Precautions for Safe Handling and Use
SECTION VIII - Control Measures
SECTION IX - Label Data
SECTION X - Transportation Data
SECTION XI - Site Specific/Reporting Information
SECTION XII - Ingredients/Identity Information

SECTION I - Material Identity

Part Number/Trade Name	WD-40 SPRAY CANS 12 OZ
National Stock Number	9150010548665
CAGE Code	09137
Part Number Indicator	A
MSDS Number	132481
HAZ Code	B

SECTION II - Manufacturer's Information

Manufacturer Name	WD-40 CO
Emergency Phone	714-275-1400

Figure 9-3   The MSDS for a WD40® bulk package.   (*continues*)

MSDS Preparer's Information

Date MSDS Prepared/Revised	PRE-HCS
Date of Technical Review	12MAR84
Active Indicator	N

Alternate Vendors

Vendor #5 CAGE	BGJJY

SECTION III - Physical/Chemical Characteristics

Hazard Storage Compatibility Code	F1-L5
NRC License Number	N/A
Net Propellant Weight (Ammo)	N/A
Appearance/Odor	LIGHT AMBER LIQ, CHARACTERISTIC ODOR
Specific Gravity	0.710
Solubility in Water	NEGLIGIBLE
Percent Volatiles by Volume	80.0
Container Pressure Code	4
Temperature Code	8
Product State Code	U

SECTION IV - Fire and Explosion Hazard Data

Flash Point Method	UNK
Lower Explosion Limit	1.80
Upper Explosion Limit	9.50
Extinguishing Media	CO*2, DRY CHEM, FOAM
Unusual Fire/Explosion Hazards	CONSIDERED EXTREMELY FLAMMABLE UNDER CONSUMER PROD SAFETY COMMISSION REG

SECTION V - Reactivity Data

Stability	YES
Materials to Avoid	STRONG OXIDIZING MATL
Hazardous Decomposition Products	NONE
Hazardous Polymerization	NO

Figure 9-3   (continued) The MSDS for a WD40® bulk package.

SECTION VI - Health Hazard Data

Symptoms of Overexposure	EYES/SKIN: IRRIT. INH: ANESTH/HDCH/DIZZ/NAUS/RESP IRRIT. ING: NAUS/DIARR/BOMIT. ASP-LUNGS: CHEM PNEUMON
Emergency/First Aid Procedures	EYES: FLUSH W/WATER FOR 15 MINS, REMOVE CONTACT LENSES IF WORN. SKIN: WASH W/SOAP & WATER, APPLY SKIN CREAM. INH: REMOVE TO RESH AIR, GIVE ARTIFICIAL RESP IF NEC, OR O*2. ING: DO NOT INDUCE VOMITING. CALL A PHYSICIAN

SECTION VII - Precautions for Safe Handling and Use

Steps if Material Released/Spilled	SPILLS UNLIKELY FROM SPRAY CANS. LEAKING CANS SHOULD BE PLACED IN PLASTIC BAG OR OPEN PAIL UNTIL PRESSURE HAS DISSIPATED, THEN TREAT AS STODDARD SOLVENT
Waste Disposal Method	EMPTY SPRAY CANS SHOULD NOT BE PUNCTURED OR INCINERATED. LIQ SHOULD BE TREATED AS STODDARD SOLVENT & DISPOSED OF IAW LOCAL, STATE, FED REG
Handling and Storage Precautions	P FROM SOURCES OF IGNITION. DO NOT TAKE INTERNALLY. AVOID EXCESSIVE INH OF SPRAY PARTICLES. DO NOT STORE ABOVE 120F.
Other Precautions	DO NOT INCINERATE OR PUNCTURE CNTNRS

Figure 9-3   (continued) The MSDS for a WD40® bulk package.

SECTION VIII - Control Measures

Respiratory Protection	NONE
Ventilation	SUFFICIENT TO KEEP SOLVENT VAPOR LESS THAN TLV
Supplemental Health/Safety Data	PROD CONTAINS PROPRIETARY CORROSION INHIBITORS & WETTING AGENTS THAT ARE INDICATED BY MFR TO BE HAZ. VAP DNSTY >1. VAP PRESS IS 50PSIG

SECTION IX - Label Data

Protect Eye	YES
Protect Skin	YES
Protect Respiratory	NO
Chronic Indicator	UNKNOWN
Contact Code	SLIGHT
Fire Code	UNKNOWN
Health Code	UNKNOWN
React Code	UNKNOWN

SECTION X - Transportation Data

Container Quantity	16
Unit of Measure	OZF

SECTION XI - Site Specific/Reporting Information

SECTION XII - Ingredients/Identity Information

Ingredient #	01
Ingredient Name	STODDARD SOLVENT
CAS Number	8052413
NIOSH Number	WJ8925000
Proprietary	NO
Percent	>50
OSHA PEL	500 PPM
ACGIH TLV	100 PPM; 9293

Figure 9-3   (continued) The MSDS for a WD40® bulk package.

SECTION XII *(continued)*	
Ingredient #	02
Ingredient Name	PETROLEUM BASE OIL
NIOSH Number	1000099PH
Proprietary	NO
Percent	>15
Ingredient #	03
Ingredient Name	LPG (LIQUEFIED PETROLEUM GAS)
CAS Number	68476857
NIOSH Number	SE7545000
Proprietary	NO
Percent	25.0
OSHA PEL	1000 PPM
ACGIH TLV	1000 PPM; 9192

Figure 9-3   (continued) The MSDS for a WD40® bulk package.

Chemical handling is a huge subject area. Handling, application, and storage of fertilizers and pesticides should be designated to a trained and licensed employee. Training in chemical, pesticide, and herbicide handling can be consolidated into the integrated pest management (IPM) concept. This is a total management approach that requires that the least possible impact on the environment be maintained. The IPM approach focuses on using the smallest amount of chemicals necessary to control the detrimental plants and pests. An additional focus is on identifying the pests first, monitoring their activity, and using the most benign method to control them. Most land grant universities work with the EPA and their state agriculture departments to educate pest management workers. At the University of Florida, the Institute of Food and Agricultural Sciences (IFAS) operates an integrated pest management center. IFAS can be found at http://ipm.ifas.ufl.edu.

There are also private schools and online courses available. The licensing needed to become an applicator requires an extensive testing and certification process. Each state has a different licensure procedure, testing, and classifications for licensing. In Florida, for instance, there are three licenses: a private license, a public license, and a commercial license. A golf course applicator must apply and hold a commercial license, not a private or public license.

RECYCLING AND DETOXIFICATION

In most states, the only acceptable way to dispose of used oil is to recycle it. Find out how to properly dispose of your waste. Even computers and other electronic devices contain hazardous materials. Find out whether your community has a hazardous waste roundup each year, and take advantage of it. In many cities, this is the best way to dispose of paint; old computers; pesticides; lead, acid, and NiCad batteries; and other waste.

If you need to neutralize a small spill, there are several ways this can be done. One product is called Spill-X,® and another is named Ampho-Mag™. Ampho-Mag™ is amphoteric, that is, it neutralizes acids and bases. Spill-X® is composed of several different bottles of chemicals, and the correct one must be selected to match the spill. Ampho-Mag™ will absorb gasoline, which makes it handy in the shop. Spill-X-S®, a different product, is for solvents, including gasoline, and both products increase the flash point of the solvent treated.

If you are dealing with a large spill, be prepared with the contact numbers of the appropriate agencies in your area. Your physical location has a dramatic influence on spill response. If you are on or near a lake, stream, or ocean, the rules are more stringent. Investigate the correct response, and build a call list of the responders you must contact in the event of a spill. It may be necessary for you to have a trained hazmat (see below) team among your employees.

This can be a burdensome necessity as the training, medical records, and record keeping for this team are detailed and rigidly prescribed.

HAZARDOUS MATERIALS RECORD KEEPING

There are a number of reports and records you must keep to satisfy state environmental protection authorities or meet the OSHA guidelines. There are two terms you will encounter with hazardous materials: *hazcom* and *hazmat*. There is a good deal of confusion about these terms. As mentioned previously, hazcom deals with hazardous chemicals and materials communications and records. Hazmat (i.e., hazardous materials) deals specifically with materials.

ACCIDENT PREVENTION

Accident prevention starts with being clean and neat. Most shop accidents are of the trip, slip, and fall variety. The first type occurs when someone trips or slips over a tool or part left on the floor. The second type occurs when someone slips on a spill that has not been cleaned up. Almost any liquid spilled on a concrete floor creates a slip hazard. Water, oil, hydraulic fluid, battery acid, and grease are common spill items. Keep a liquid-absorbent compound, such as Oil-Dri™, nearby to absorb spills. The manufacturers of these compounds also can supply emergency spill kits, absorbent mats, socks, pillows, and rolls for larger spills. If you have a state department of environmental protection, it can provide you with the spill kit requirements for your location.

Lifting injuries are a close second in the shop accident race. Most people have never been taught how to lift correctly. There are basic techniques that need to be taught that protect your back. In many heavy industries in which back injuries were once prevalent, hand lifting has been eliminated. Lift tables, small cranes, hydraulic pallet jacks, and other specialized equipment are used exclusively. Some employers with excessive health and disability insurance claims resulting from improper lifting have driven employers and insurance companies to make lifting by hand a firing offense.

Fire prevention is another area in which careful floor plan designs and equipment locations can greatly reduce risk. By locating the spark-producing equipment intelligently, the risk of fire can be minimized. Grinding zones, welding and fabrication zones, battery-charging zones, and other zones in which a spark from one can lead to a fire in the other need to be physically separated. Fire laws vary from state to state, but usually 35 feet is the minimum separation distance between these zones.

The number and type of fire extinguishers are usually dictated by the local or state fire marshal. The most expedient way to guarantee compliance is to hire a specialty contractor who will provide turnkey service on your extinguishers. The contractor will survey your facility and submit a bid for installation and maintenance on your extinguishers. There may be several companies in the area that provide this service, so you can obtain bids. Check each vendor's license and performance record with the fire marshal's office.

ACCIDENT REPORTS AND REPORTING

Most golf courses, whether public or private, fall under the Standard Industry Code 799. This means they are required to follow OSHA accident reporting procedures. You may be partially exempt from some record keeping if you have ten or fewer employees. The general rule is that if there is an accident resulting in a death or if three or more people are injured, OSHA must be called in to investigate. The best course of action is to call the closest OSHA office and ask about your specific situation. OSHA office locations and their telephone numbers are located by state and region on their website at www.osha.gov. OSHA also has a project called the Safety and Health Achievement Recognition Program (SHARP) that may be of help to your facility. You call OSHA for an inspection, they make recommendations, and you remediate any deficiencies. If you become SHARP qualified, you receive an exemption from inspections for the length of your certification.

Some facilities have found it more expedient to hire a private-sector OSHA consulting contractor to create a plan for compliance. This is a potential way to stay in compliance without becoming an OSHA "expert." There are contractors listed in almost any telephone book. As you would do with any contractor, be sure to investigate the contractors' past performance and credentials carefully.

As you would expect, ignorance of the law is no excuse. If you do not have OSHA plans in place, complete these immediately. Study the laws, talk to your peers, and hire a contractor if necessary. OSHA has a history of steep fines for noncompliance, but facilities that take a proactive approach with them usually receive generous treatment.

FIRST AID AND FIRST RESPONDERS

One of the best guides to first aid in the workplace can be found at www.osha.gov/Publications/OSHA3317first-aid.pdf. This guide is an excellent overview of current best practices in workplace first aid. The first page of this guide states, "The guide is advisory in nature, informational in content, and is intended to assist employers in providing a safe and healthful workplace."

This guide is easy to follow and lists all the basic preparations that need to be made to have an effective first aid

Figure 9-4   This OSHA guide, "Best Practices Guide: Fundamentals of a workplace First-Aid Program," is an excellent overview of best workplace first aid.

(continues)

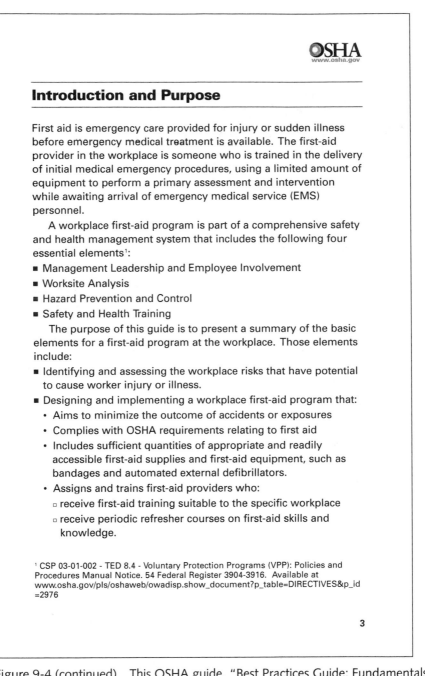

OSHA
www.osha.gov

Introduction and Purpose

First aid is emergency care provided for injury or sudden illness before emergency medical treatment is available. The first-aid provider in the workplace is someone who is trained in the delivery of initial medical emergency procedures, using a limited amount of equipment to perform a primary assessment and intervention while awaiting arrival of emergency medical service (EMS) personnel.

A workplace first-aid program is part of a comprehensive safety and health management system that includes the following four essential elements[1]:

- Management Leadership and Employee Involvement
- Worksite Analysis
- Hazard Prevention and Control
- Safety and Health Training

The purpose of this guide is to present a summary of the basic elements for a first-aid program at the workplace. Those elements include:

- Identifying and assessing the workplace risks that have potential to cause worker injury or illness.
- Designing and implementing a workplace first-aid program that:
  - Aims to minimize the outcome of accidents or exposures
  - Complies with OSHA requirements relating to first aid
  - Includes sufficient quantities of appropriate and readily accessible first-aid supplies and first-aid equipment, such as bandages and automated external defibrillators.
  - Assigns and trains first-aid providers who:
    - receive first-aid training suitable to the specific workplace
    - receive periodic refresher courses on first-aid skills and knowledge.

[1] CSP 03-01-002 - TED 8.4 - Voluntary Protection Programs (VPP): Policies and Procedures Manual Notice. 54 Federal Register 3904-3916. Available at www.osha.gov/pls/oshaweb/owadisp.show_document?p_table=DIRECTIVES&p_id =2976

3

Figure 9-4 (continued)   This OSHA guide, "Best Practices Guide: Fundamentals of a workplace First-Aid Program," is an excellent overview of best workplace first aid.

OSHA
www.osha.gov

- Instructing all workers about the first-aid program, including what workers should do if a coworker is injured or ill. Putting the policies and program in writing is recommended to implement this and other program elements.
- Providing for scheduled evaluation and changing of the first-aid program to keep the program current and applicable to emerging risks in the workplace, including regular assessment of the adequacy of the first-aid training course.

   This guide also includes an outline of the essential elements of safe and effective first-aid training for the workplace as guidance to institutions teaching first-aid courses and to the consumers of these courses.

The Risks: Injuries, Illnesses and Fatalities

There were 5,703 work-related fatalities in private industry in 2004. In that same year there were 4.3 million total workplace injuries and illnesses, of which 1.3 million resulted in days away from work.

   Occupational illnesses, injuries and fatalities in 2004 cost the United States' economy $142.2 billion, according to National Safety Council estimates. The average cost per occupational fatality in 2004 exceeded one million dollars. To cover the costs to employers from workplace injuries, it has been calculated that each and every employee in this country would have had to generate $1,010 in revenue in 2004.[2]

   Sudden cardiac arrest (SCA) may occur at work. According to recent statistics from the American Heart Association, there are 250,000 out-of-hospital SCAs annually. The actual number of SCAs that happen at work are unknown. If an employee collapses without warning and is not attended to promptly and effectively, the employee may die. Sudden cardiac arrest is caused by abnormal, uncoordinated beating of the heart or loss of the heartbeat altogether, usually as a result of a heart attack.

[2] National Safety Council. (2006). *Injury Facts,* 2004 - 2006 Edition. Itasca, IL, p. 51.

4

Figure 9-4 (continued)   This OSHA guide, "Best Practices Guide: Fundamentals of a workplace First-Aid Program," is an excellent overview of best workplace first aid.

program (see Figure 9-4). The guide has segments on risk assessment, designing a first aid program, OSHA requirements, first aid supplies, defibrillators, first aid courses, and training. This booklet states that every facility has need of a first aid kit or kits. There is also information on the OSHA General Industry Standards, Medical Services and First Aid, 29 CFR 1910.151, Subpart K, which states:

(a) The employer shall ensure the ready availability of medical personnel for advice and consultation on matters of plant health.

(b) In the absence of an infirmary, clinic, or hospital in near proximity to the workplace which is used for the treatment of all injured employees, a person or persons shall be adequately trained to render first aid. Adequate first aid supplies shall be readily available.

(c) Where the eyes or body of any person may be exposed to injurious corrosive materials, suitable facilities for quick drenching or flushing of the eyes and body shall be provided within the work area for immediate emergency use. (Revised June 18, 1998.)

Section c requires an eyewash station in each hazardous area. This means that any area that has potential exposure, such as the shop area and the chemical storage area, needs eyewash stations. This booklet suggests finding out how long it will take the local EMS to respond to your location. Additionally, OSHA requires a first aid–trained first responder if the facility does not have an emergency room, clinic, or hospital nearby.

Your first aid kit(s) should meet the ANSI 308.1-2003 standards. This sounds complicated, but all first aid supply companies are familiar with the standard, have kits to meet the standard, and will ask you how many employees you have so they can assess the number of kits you will need. Most of these companies will contract with you to keep your kits refilled and up to date.

chapter

10

Conclusions

This book was written with the desire to help facility managers, equipment managers, and anyone who has to design, build, or manage a turf equipment maintenance facility to be successful. There are many areas that must be mastered to become an accomplished turf equipment facility manager. A universalistic personality is necessary. Understanding what you do well and what you do not is essential. Willingness to delegate and the ability to select, train, and manage your personnel are top priorities. If you are a superintendent or a sports turf manager or are in a municipal environment, you must select the best possible assistant and equipment manager. You will likely be more comfortable choosing an assistant. Your skill and expertise are obviously in the same areas required of a good assistant. If you do not have a good deal of mechanical expertise and are unfamiliar with the workings of an efficient shop, you will have a more difficult time selecting an equipment manager. Equipment managers and technicians are in short supply; the market has been tight for nearly ten years. Too often, the result is desperation and bad hiring choices.

As of 2008, there continues to be a national shortage of turf equipment technicians. The job market is strong with excellent salaries, but colleges like Lake City Community College (Florida) and others have a difficult time filling the turf equipment classes to meet the local demand, let alone satisfy the national demand for technicians. There does not appear to be an easy solution to this problem.

Why are turf, municipal, and sports field managers having trouble finding technicians? The trainee and student pipeline is not filling. In general, high school students are poorly informed about most of the technical careers that exist. If you talk to hiring managers in related industries, you will learn that there is also a national shortage of plumbers, electricians, carpenters, automobile mechanics, landscape managers, irrigation technicians, skilled workers for the logistics and supply chain industries, and an array of other similar careers. Most of these skilled technical careers are overlooked by students, parents, and counselors. Unfortunately, a career as a turf equipment manager or technician also falls into this category. One university-level professional recruiter explained, tongue in cheek, that the turf equipment technician field is a "discovery" career. He said that the only way a student would choose this career field is if he or she accidentally discovered it. This lack of career awareness, which was discussed in Chapter 8, is a huge issue that will not be resolved until turf industry leaders make an effort at the local level to inform middle and high school students about this career.

The authors do not see any change in the current situation unless the affected parties in the turf industry at the state and national levels decide to make a concerted effort to tackle this problem. There are public technical schools and community colleges willing to train turf equipment managers, but the schools cannot help without increased enrollment.

Golf course superintendents and sports turf managers may be able to fill a few turf equipment positions by hiring local people with some mechanical interest or by stealing trained technicians away from other golf courses, automobile dealerships, etc., but this is a dead-end solution. As the turf equipment industry has become more specialized, the situation has become more difficult. This points to the increasing need for students to be aware of and trained for the specialty career of turf equipment manager. Whether this happens or not is in the hands of golf course superintendents, equipment manufacturers, distributors, and other turf managers across the country. As long as the industry is satisfied with the turf equipment managers it is able to find now, there is little hope for change.

Educate yourself; read the chapters in this book that focus on the shop, the technician, and where to find one. The more you understand about this possibly unfamiliar territory, the more comfortable you will become with interviewing and hiring. Explore your job market. Do realize that the market is tight, but with the right incentives, good technicians and equipment managers can be found.

We wish you success and great management.

Appendix

Equipment and Engine Manufacturers

Briggs and Stratton®, Commercial Engines
www.commercialpower.com

Club Car®, Golf Cars
www.clubcar.com

DC Atlas Co., AgriMetal greens roller
www.dcatlas.com/turf-greensroller.htm

Echo® Outdoor Power Equipment, Handheld Power Equipment
www.echo-usa.com

E-Z-GO®, Golf Cars
1-800-241-5855
www.ezgo.com

Gravely®, Rotary Mowers
www.gravely.com

Honda, Commercial Engines
www.honda-engines.com

Husqvarna, Handheld Power Equipment
www.usa.husqvarna.com

Jacobsen®, Golf and Turf Equipment
www.jacobsen.com

John Deere®, Golf and Turf Equipment
www.deere.com/en_US/golfturf/index.html?location

Kawasaki®, Commercial Engines and Handheld Power
Equipment
www.kawpowr.com/powerproducts.asp

Kohler®, Commercial Engines
www.kohlerengines.com/index.jsp

Kubota®, Engines, Commercial Engines
www.kubotaengine.com

Kubota®, Tractors, Mowers and Utility Vehicles
www.kubota.com/f/home/home.cfm

Red Max®, Komatsu Zenoah, Handheld Power Equipment
www.redmax.com

Salsco, Greens Roller
www.salsco.com
1-800-872-5726

Shindaiwa, Handheld Power Equipment
www.shindaiwa.com/nam/en/products/index.php

Smithco, Golf and Turf Equipment
1-877-833-7648
www.smithco.com

Speed Roller®, Diversified Manufacturing Inc., Greens Roller
www.speedroller.com

STIHL® Incorporated, Handheld Power Equipment
www.stihl.usa.com

Subaru Robin, Commercial Engines
www.subarupower.com

Toro® Company, Golf and Turf Equipment
www.toro.com/golf/index.html

Tru-Turf, greens roller
www.truturf.com

Vicon Spreaders, Fertilizer Spreaders
www.kvernelandgroup.com/welcome

Yamaha Golf-Car Company, Golf Cars
www.yamahagolfcar.com

Fuel, Oil and Storage

ConVault® double-walled storage tanks.
www.convault.com

Double-wall steel tanks.
www.bhtank.com/double_wall.asp#a_strip

EPA used oil guide.
www.epa.gov/epaoswer/hazwaste/usedoil/usedoil.htm#filters
The EPA oil spill management hotline can be reached
through the National Response Center at 1-800-424-8802 or

Management of aboveground storage tanks.
www.epa.gov/safewater/sourcewater/pubs/ast.pdf

Oil filter disposal hotline.
Filters Manufacturing Council 1-800-99FILTER

1-202-260-2342, or online at http://www.nrc.uscg.mil/nrchp.
html.

Steel, double-walled tanks with secondary containment.
www.ipstanks.com/fguard.htm

Preventive Maintenance and Management Software

This site focuses primarily on trucking fleet maintenance
software vendors. It lists several companies that offer global
positioning system (GPS) tracking of unit locations. This
may be a future need for the turf equipment manager.
www.capterra.com/landing/vfm

Datamasters has a program called Shop Man that does basic
shop management tasks but seems to lack true preventive
maintenance (PM) capability.
www.datamasters.net/products.html

Link It Software produces very heavy-duty equipment
and fleet maintenance software. Its customers include Los
Alamos Laboratories and Walt Disney World/Epcot.
www.ez-maintenance.com

This is the website for the MagnaTag® magnetic board visible maintenance system.
www.magnatag.com/page/MF/board/maintenance-schedule-
board/

ManagerPlus is a basic PM software program. This program
will also track fuel, chemicals, and parts.
www.managerplus.com

This company produces Maintenance Pro and Fleet Maintenance Pro. It seems to be focused primarily on heavy equipment PM and is reasonably priced.
www.mtcpro.com/comparis.htm

The New Standards Institute is primarily focused on industrial maintenance but has a free software program for calculating parts reorder points.
www.newstandardinstitute.com/catalog_reftools.cfm

A very basic, easy-to-use program called Turf Log. This program does not appear to have PM capability.
www.precision-data-services.com

TRIMS is focused specifically on the turf management/turf equipment management industry. This is a wide-ranging and versatile program that is a near industry standard.
www.trims.com/mindex.htm

Regulation and OSHA

Databases and software from the Environmental Protection Agency (EPA).
www.epa.gov/epahome/Data.html

Keller's Official OSHA Safety Handbook
This is an easy-to-understand handbook of OSHA regulations written for employees.
www.jjkeller.com

Database of material safety data sheets.
www.msdssearch.com

OSHA database of chemicals.
www.osha.gov/web/dep/chemicaldata/#target

Reel Grinders, Bed Knife Grinders, and Accessories

Bernhard and Company, Express Dual, Anglemaster Grinders
1-888-474-6348
www.expressdual.com

Foley United, ACCU-Pro, ACCU-Sharp, ACCU-Master Grinders
1-800-225-9810
www.foleyunited.com

Neary Technologies, Grinders
1-800-233-4973
www.nearytec.com

Simplex Ideal Peerless™, Grinders
1-800-888-6658
www.sipgrinder.com

Turf Pride LLC, PowerEdge Reel Sharpener
1-888-427-7605
www.turfprideusa.com/products-reelsharp.html

Storage and Shop Benches

Eagle Manufacturing Company, Flammable Storage Cabinets
www.eagle-mfg.com/flammable.html

Knaack®, Tool Storage and Workbenches
www. www.jobsitestoragebox.com

Lyon® All-Welded Storage, Parts Storage Cabinets
www.lyon-cabinets.com/Lyon-Plastic-Bin-Storage-Cabinets-
Sale.htm

Strong Hold, Workbenches
1-800-880-2625
www.strong-hold.com

Tools, Lifts, and Shop Equipment

Accuproducts International, Height of Cut Gauges
www.accuproducts.com

Ammco® Coats®, Hennessey Industries, Lifts and Tire Changers
1-800-688-6359
www.ammcoats.com

Autobarn, Tools and Tire Gauges
www.autobarn.net/the-tool-chest.html

Ben Pearson Tubemaster, Lifts
1-800-436-1327
www.ben-pearson.com

Cornwell Tools
www.cornwelltools.com/index.asp

Deltran Corporation, Battery Tender®, Automatic Battery Chargers
www.batterytender.com

Golf Lift®, Derek Weaver Company, Lifts
1-800-788-9789
www.golf-lift.com

Heftee Industries, Lifts
1-800-755-7540
www.heftee.com

The Home Depot®, Tools
www.homedepot.com

Ingersoll Rand, Rotary Air Compressors
air.ingersollrand.com/IS/product.aspx-en-12858

Jack's Small Engines, Tools
www.jackssmallengines.com/help.cfm

Lowe's®, Tools
www.lowes.com

Mac Tools
https://www.mactools.com/portal/site/mactools

Magma-Matic Corporation, Blade Balancer, Track Gauge
www.magna-matic.com

Manitowoc Lifts
www.manitowoclifts.com/manitowoc/turf

Matco, Tools
www.matcotools.com/index.jsp

Mohawk Lifts
www.mohawklifts.com/consumer/turf.php

Northern Tool, Tools, Compressors, and Replacement Engines
www2.northerntool.com

PMW Precision Metal Works, Lifts
www.pmwequipment.com/mp6000.htm

R&R Products, Shop Tools and Shop Supplies
1-800-528-3446
www.rrproducts.com

Rotary Lift®
www.rotarylift.com/library/webPost/marCom_Materials/
22.pdf

Sears, Air Compressors, Hand Tools, Pneumatic Tools
www.sears.com

Snap-on Tools
www.snapon.com/tools/hand-tools.asp

Superior Car Care, Accutire® Tire Gauges
www.superiorcarcare.net/actiprgu.html

The Tool Warehouse, Shop Tools and Battery Chargers
www.thetoolwarehouse.net

Trion® Lifts
1-800-426-3634
www.trionlifts.com

W.W. Grainger, Inc., Air Compressors
www.grainger.com/production/info/air-compressor.htm

About the Authors

JOHN R. PIERSOL

John R. Piersol has a Bachelor of Science (BS) degree in Plant Science from the University of Delaware and a Master of Science degree in Horticulture from Colorado State University. After receiving his BS degree in 1970, he spent 21 months of active duty in the Naval Air Reserve with Antarctic Development Squadron Six, which offered logistics support to the United States Antarctic Research Project. He accepted a position as the landscape instructor at Lake City (Florida) Community College in 1974, and in 1987, he became chairman of a division that now includes the following programs: AS degree programs in Golf Course Operations, Landscape Technology, and Agribusiness Management, and certificate programs in Forest Operations, Irrigation Technology, Turf Equipment Technology, Horticulture, and Pest Control Operations. Mr. Piersol has presented seminars on golf course, landscape, and education topics throughout Florida, the United States, and internationally. He can be reached at Lake City Community College as follows:

John R. Piersol, Chair
Director of Golf/Landscape/Forestry
Lake City Community College
149 SE College Place
Lake City, FL 32025
Phone: 386-754-4225
Email: piersolj@lakecitycc.edu
Website: www.lakecitycc.edu

HARRY V. SMITH

Harry V. Smith has more than 25 years of experience in the turf equipment and allied industries as an equipment dealer, distributor's territory manager, turf equipment management professor, and technical trainer. Now retired, Mr. Smith was a Professor and Program Coordinator of the Turf Equipment Management Program at Lake City Community College (Florida), where he taught basic electrical theory, single- and multicylinder gasoline and diesel engine diagnostics, preventive maintenance of turf equipment, maintenance shop organization, budgeting, and golf course maintenance shop management. Mr. Smith received his Master of Science in Education in Human Resource Development, Adult Training and Development from Georgia State University in Atlanta, Georgia. He is an ASE-certified Master Automobile Technician and a certified Briggs and Stratton Master Service Technician.

Index